Mastering Mechanical Desktop: Surface Modeling

Mastering Mechanical Desktop: Surface Modeling

RON K. C. CHENG

The Hong Kong Polytechnic University

Autodesk.
Press

I(T)P   An International Thomson Publishing Company

Albany • Bonn • Boston • Cincinnati • Detroit • London • Madrid
Melbourne • Mexico City • New York • Pacific Grove • Paris • San Francisco
Singapore • Tokyo • Toronto • Washington

Trademarks

AutoCAD, Mechanical Desktop, AutoCAD Designer, AutoSurf, AutoVision, and 3D Studio are a registered trademarks of Autodesk, Inc.
The ITP logo is a trademark under license.
Windows is a registered trademark of Microsoft Corporation.

Copyright ©1998
PWS Publishing Company
Autodesk Press imprint
an International Thomson Publishing Company

Printed and bound in the United States of America.
1 2 3 4 5 6 7 8 9 10 — 01 00 99 98 97

For more information contact:

Autodesk Press
3 Columbia Circle, Box 15-015
Albany, New York 12212-5015

International Thomson Publishing Europe
Berkshire House
168–173 High Holborn
London WC1V 7AA
England

Thomas Nelson Australia
102 Dodds Street
South Melbourne, 3205
Victoria, Australia

Nelson Canada
1120 Birchmont Road
Scarborough, Ontario
Canada M1K 5G4

International Thomson Publishing Southern Africa
Building 18, Constantia Park
240 Old Pretoria Road
P.O. Box 2459
Halfway House, 1685 South Africa

International Thomson Editores
Campos Eliseos 385, Piso 7
Col. Polanco
11560 Mexico D.F., Mexico

International Thomson Publishing GmbH
Königswinterer Strasse 418
53227 Bonn, Germany

International Thomson Publishing France
Tour Maine-Montparnasse
33, Avenue du Maine
75755 Paris Cedex 15, France

International Thomson Publishing Asia
221 Henderson Road
#05-10 Henderson Building
Singapore 0315

International Thomson Publishing Japan
Hirakawacho Kyowa Building, 31
2-2-1 Hirakawacho
Chiyoda-ku, Tokyo 102
Japan

Assistant Editor: Suzanne Jeans
Production Editor: Andrea Goldman
Manufacturing Buyer: Andrew Christensen
Marketing Manager: Nathan Wilbur
Interior Design/Cover Image: Ron C. K. Cheng
Cover Design: Autodesk, Inc.
Cover Printer: Phoenix Color Corp.
Printer & Binder: Courier–Westford

Library of Congress Cataloging-in-Publication Data

Cheng, Ron.
    Mastering Mechanical Desktop: surface modeling / Ron K. C. Cheng.
        p.   cm.
    Includes index.
    ISBN 0–534–95085–X
    1. Engineering graphics. 2. Autodesk Mechanical desktop. 3. Engineering design--Data processing. 4. AutoSurf. I. Title.
T353.C519 1997   97–29789
620'.0042'02855369--dc21   CIP

Contents

Preface

This book is intended for people who would like to use AutoSurf to produce and manufacture engineering designs. AutoSurf is a component of Mechanical Desktop. (The other component is AutoCAD Designer, which is covered in another book in this series.)

AutoSurf is a Non-Uniform Rational B-Spline (NURBS) surface modeling system, which allows you to use NURBS mathematics to create NURBS surfaces and free-form surfaces. Because Mechanical Desktop is a member of the AutoCAD family, objects created in AutoSurf are fully compatible with AutoCAD Designer solids and AutoCAD native solids. You can use AutoSurf surfaces to cut these solids and incorporate free-form surface features in them. You can also convert an AutoCAD Designer solid or an AutoCAD native solid to a set of AutoSurf NURBS surfaces.

This book will show you how to use AutoSurf to create 3D wireframes from which you can make various kinds of surfaces, compose 3D surface models, prepare engineering documents from the surface models, and inter-operate AutoSurf surfaces with AutoCAD Designer solids and AutoCAD native solids.

This book consists of six chapters and an appendix. Chapter 1 introduces the AutoSurf application. Chapter 2 guides you to produce a set of wireframes. Chapter 3 leads you to compose a surface model using the wireframes you prepared in Chapter 2. Chapter 4 provides additional practice of the skills covered in Chapters 2 and 3. Because AutoSurf is compatible with AutoCAD Designer and native solids, you will learn to inter-operate the three kinds of objects in Chapter 5. Chapter 6 shows you how to output an engineering drawing. The AutoSurf commands and variables used are summarized in the appendix. After working through the projects in this book, you should be well on you way to mastering the AutoSurf techniques used to produce engineering designs.

Acknowledgments

This book never would have been realized without the contributions of many individuals.

I am grateful to the following reviewers for their thoughtful suggestions and help:

- Robert A. Chin, Department of Industrial Technology, East Carolina University
- Hollis Driskell, Department of Drafting and Design, Trinity Valley Community College
- Michael Stewart, Department of Engineering Technology, University of Arkansas
- Ed Wheeler, Engineering Department, University of Tennessee at Martin

Several people at PWS Publishing also deserve special mention, particularly Bill Barter, Jonathan Plant, Suzanne Jeans, Andrea Goldman, Tricia Kelly, and Monica Block.

Chapter 1

Introduction

AutoSurf is a surface modeling system. It is a useful tool for designing and creating free-form surfaces and surface models in a computer. We can use the surface modeling tool in a number of ways. At one end of the spectrum, we can use the electronic data to control a computerized numerical control machine to manufacture the surface profile of the product. At the other end of the spectrum, we can use the data to produce photo-realistic rendering on a computer.

If we look around us, we can easily find many objects with free-form surfaces. Some examples are the handle of a shaver, the casing of a computer pointing device, the casing of a mobile phone, and the body panels of an automobile.

AutoSurf is a Non-Uniform Rational B-Spline (NURBS) surface modeling system for design and manufacturing. It uses Non-Uniform Rational B-Spline mathematics for the creation of splines and surfaces. It also offers additional sophisticated tools for creating and editing wireframes for subsequent generation of NURBS surface models. Using its import utilities, you can create 3D surfaces from IGES files or digitized data. Given the shell thickness, you can evaluate mass properties from a surface model.

To add flexibility in model creation, you can use a NURBS surface to cut an AutoCAD native solid or an AutoCAD Designer parametric solid. Working in the opposite direction, you can convert an AutoCAD native solid or an AutoCAD Designer solid to become a set of NURBS surfaces.

In addition to being an advanced manufacturing tool for the creation of 3D wireframe and NURBS surfaces, AutoSurf provides geometric creation tools to generate detailed drawings from a 3D surface model. You can also render a surface model in AutoCAD or use a surface model in AutoVision and 3D Studio to produce photo-realistic rendering and animation.

1.1   NURBS Splines and Surfaces

A surface in a computer is a mathematical expression to represent a three-dimensional shape with no thickness. NURBS mathematics is the most advanced tool in surface modeling. It allows the implementation of multi-patch surfaces with cubic surface mathematics and maintains full continuity control even with trimmed surfaces.

AutoSurf employs NURBS mathematics for defining curves and surfaces. Basically, AutoSurf uses NURBS splines to create NURBS surfaces. However, it also accepts 3D polylines. When you input a 3D polyline, AutoSurf will fit it to become a spline according to a set of rules before using them to create a surface.

1

By using AutoSurf, you can create three kinds of surfaces: primitive surfaces, free-form surfaces, and derived surfaces. In addition, you can trim a surface to obtain a boundary of a particular shape.

1.2   Primitive Surfaces

As the name implies, primitive surfaces are basic geometric shapes. They include conical surface, cylindrical surface, spherical surface, torus surface, and planar surface. To produce a primitive surface, you do not need to prepare any construction lines or wireframes. All you need to do is to specify a geometric shape and state the dimensions and location. For example, if you want to produce a cylindrical surface, you need to state only the location of the center of the base, the diameter, and the height. These primitive surfaces, although easy to create, have very limited use because there are not too many applications of them in engineering design. Figure 1.1 shows five examples of primitive surfaces.

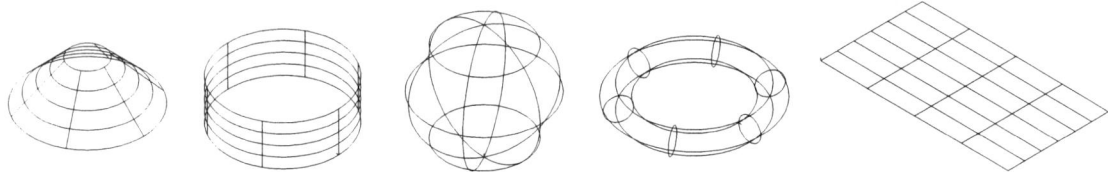

Figure 1.1 Primitive surfaces

1.3   Free-Form Surfaces

The second surface type, free-form surfaces, is the most important. These are the surfaces that you will use most frequently in your design. This type includes revolved surface, extruded surface, ruled surface, lofted U surface, swept surface, lofted UV surface, and tubular surface. Except for the tubular surface, the starting point to produce a free-form surface is to create a set of control curves that define the profiles and contours of the surface. The computer then computes and creates a set of surface data based on the curves.

In general, a free-form surface needs two sets of wireframes in two orthogonal directions to define the control curves. To distinguish them from the X axis and Y axis terminology, the two directions are called U direction and V direction. Wireframes in these directions are called U-lines and V-lines respectively.

The simplest types of free-form surface are the revolved surface and the extruded surface. These surfaces are simple because they need only a single wireframe to define their shapes. For a revolved surface, you need to produce a wireframe that defines the cross section along the axis of revolution. To make the surface, you specify the section wireframe and an axis of rotation. Figure 1.2 shows a wireframe rotated about an axis of 180°.

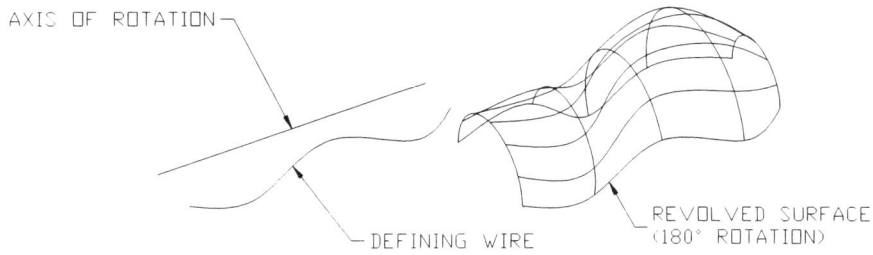

Figure 1.2 Revolved surface

Similar to a revolved surface, an extruded surface also needs a single wireframe to define its cross section. To make the extruded surface, you specify the section wireframe, the direction of extrusion, and the taper angle. See Figure 1.3.

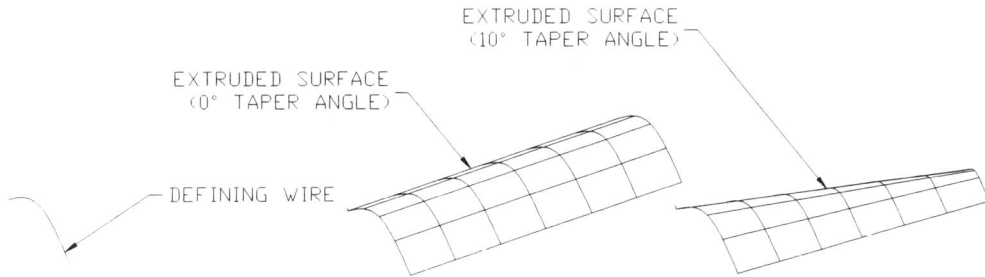

Figure 1.3 Extruded surfaces

With two wireframes, you can create a surface with a changing cross section — a ruled surface. To produce a ruled surface, you need to specify two input wireframes, which are called U-lines. The resulting surface will change in cross section linearly from the first wireframe to the second wireframe in one direction. In the other direction, the cross section is a straight line. Figure 1.4 shows a ruled surface.

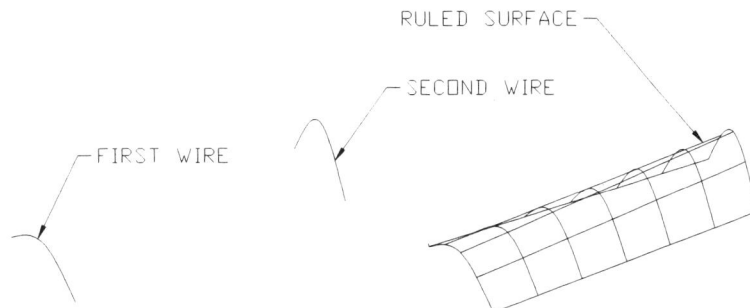

Figure 1.4 Ruled surface

To make a surface that has a variable cross section, you use three or more U-lines as input wireframes. The resulting surface is called a lofted U surface. Its cross section in the U direction will change smoothly from the first wireframe to the second wireframe, and then from the second wireframe to the third wireframe. If you have a fourth wireframe, the cross section changes from the third wireframe to the fourth wireframe as well. As a result, the cross section in the V direction also changes smoothly from one edge to the other edge. Figure 1.5 shows a lofted U surface created from three U-lines.

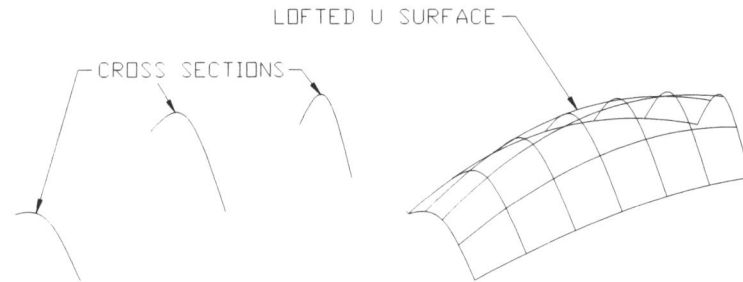

Figure 1.5 Lofted U surface

As you can see in Figure 1.5, the V direction cross sections of a lofted U surface are smooth spline curves that pass through the U-lines. Their shapes are determined by how the U-wires change from one section to another. If you want to exercise some control over the contours of the surface in the V direction, then you can specify one or two rails for the U-lines to transit. Such a surface is called a swept surface. Figure 1.6 shows a swept surface with a single rail.

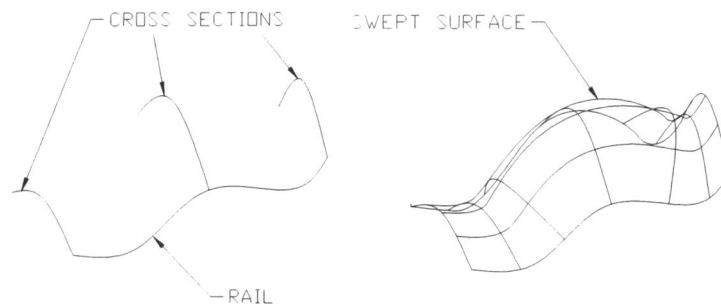

Figure 1.6 Swept surface with a single rail

If you compare Figure 1.6 with Figure 1.5, you can see how the rail controls the transition of the U-lines as they change from one section to another. In Figure 1.6, the rail controls one end of the U-lines. If you want to control both ends of the U-lines, you should use two rails. Figure 1.7 shows a swept surface with two rails.

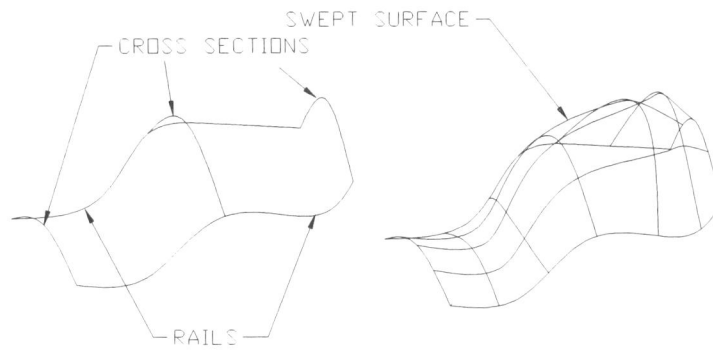

Figure 1.7 Swept surface with two rails

When you use two rails instead of one in making a swept surface, the U-line cross sections change in two ways. First, they change from one section to another. Second, they deform in shape according to the distance between the rails. Compare Figure 1.7 with Figure 1.6 to see the difference.

Although the use of rail controls how the U-lines transit, the cross sections in the V direction are not dominated by the rails. They are simple splines going from one section to another section. If you want to have full control over the V direction cross sections, then you have to specify two sets of lines -- one set of U-lines and one set of V-lines. The resulting surface is called a lofted UV surface. See Figure 1.8.

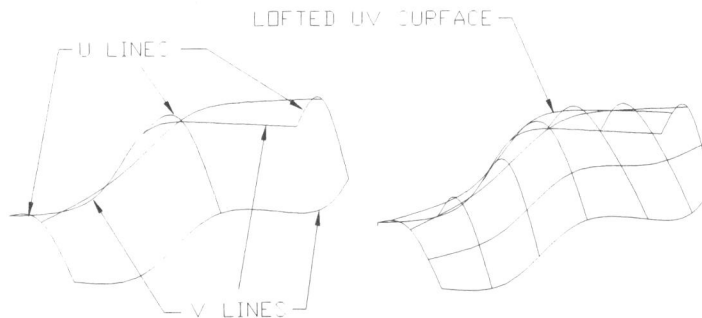

Figure 1.8 Lofted UV surface

At a glance, the lofted UV surface looks very similar to the swept surface with two rails. However, they are different. In a swept surface, the amount of control of the rails over the surface is limited only to the transition of the U-lines. In a lofted UV surface, the U-lines control the surface in the U direction and the V-lines control the surface in the V direction.

The last kind of free-form surface is the tubular surface. This kind of surface is entirely different from the aforementioned surfaces. It needs only a 3D polyline with a number of straight line segments. In making the surface, you have to specify the diameter of the tube and the radius of the bends. See Figure 1.9.

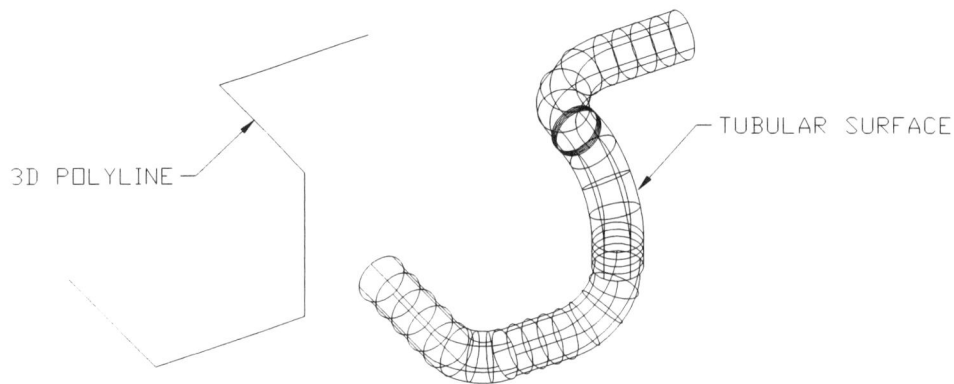

Figure 1.9 Tubular surface

With the exception of the tubular surface, all free-form surfaces have one thing in common: They all need to be created from smooth wireframes. How to construct a set of proper 3D wireframes is perhaps the first difficulty that you might encounter in surface modeling. In the process of designing, you usually start by thinking about the surface but not the wireframes. If you want to create the surface in a computer, you need to be able to perceive such defining wireframes from a given surface. From these wireframes, you can decide which type of surface it is, and then produce the surface. In the following chapters, you will be guided to construct 3D wireframes as well as 3D surfaces. In pursuing the guided tutorials, you should try to form a mental model to relate the 3D wireframes to the 3D surfaces. It is to be hoped that you can do a reverse process of seeing the wireframes when a surface is given.

1.4   Derived Surfaces

A surface model is a set of surfaces put together to enclose a volume to give the shape and appearance of the object that the model represents. To create a surface model, you have to create a number of surfaces. With two or more surfaces, you need to do something at their joining edge. To this, you can use a fillet surface, corner surface, or blended surface. These three types of surfaces, together with offset surface, are known collectively as derived surfaces.

A fillet surface treats the edge between two intersecting surfaces by providing a surface with a circular cross section. Figure 1.10 shows a fillet surface of constant radius formed between two planar surfaces.

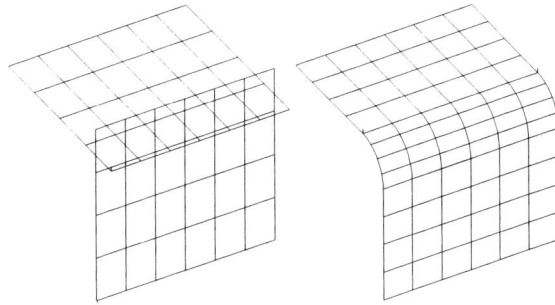

Figure 1.10 Constant radius fillet surface

In making a fillet surface, you can set the fillet radius to vary linearly or cubically. In a linear variable fillet surface, the fillet radius changes linearly from one set value to another set value. In a cubical variable fillet surface, the fillet radius changes cubically from one to another. See Figure 1.11.

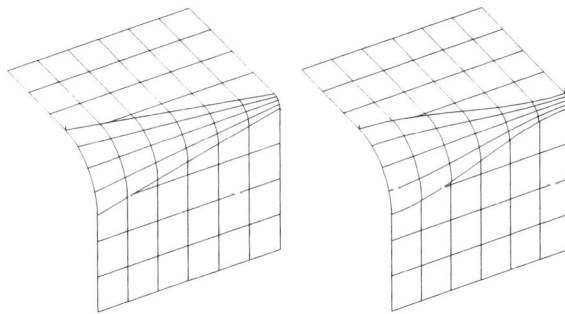

Figure 1.11 Linear variable fillet (left) and cubical variable fillet (right)

A fillet surface is used to treat two intersecting surfaces. If you have three intersecting surfaces to treat, you can first form fillets in pairs, and then treat the intersecting fillets with a corner fillet. See Figure 1.12.

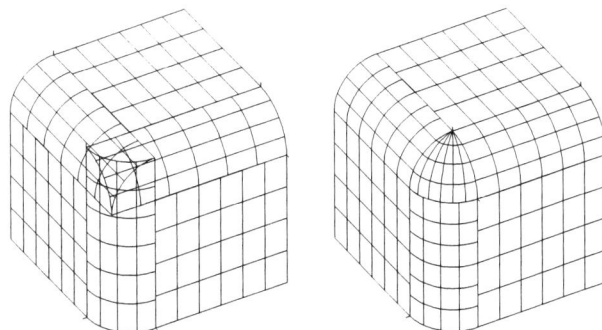

Figure 1.12 Intersecting fillet surfaces (left) and corner fillet surface (right)

In a design, two adjacent surfaces do not always have to intersect each other. If you come across two nonintersecting surfaces, you can fill in the gap between them by blending. Figure 1.13 shows a blended surface formed between two nonintersecting surfaces.

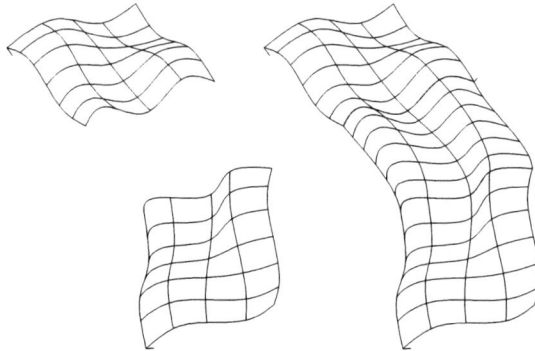

Figure 1.13 Nonintersecting surfaces (left) and blended surface formed

In addition to blending two surfaces, you can blend three or four surfaces. See Figure 1.14 and Figure 1.15.

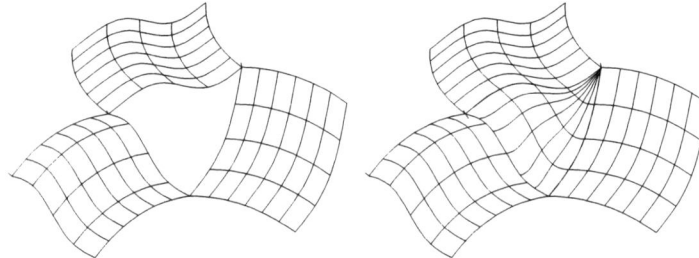

Figure 1.14 Blended surface among three edges

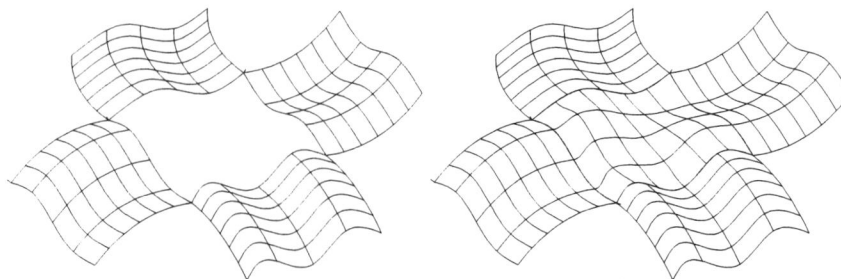

Figure 1.15 Blended surface among four edges

In many designs, a second surface is needed to run at a constant distance away from another surface. To make such a second surface, you do not have to prepare a set of offset wireframes. Instead, you can simply derive an offset surface from an existing surface.

Figure 1.16 shows an example of an offset surface applied to produce the earpiece of a walkie talkie.

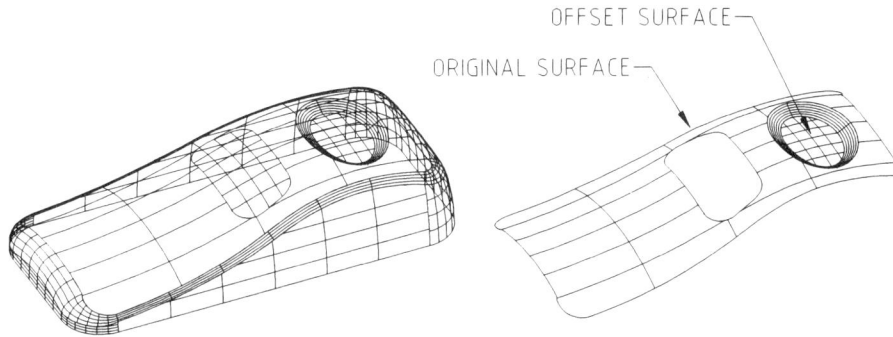

Figure 1.16 Offset surface example

1.5 Trimmed Surfaces

To produce a smooth surface, it is necessary to use smooth defining wireframes and smooth boundary lines. However, most of the surfaces that we use in design do not necessarily have a smooth boundary, although they have smooth profiles.

Figure 1.17 shows the surface model of an automobile body panel. This is a smooth surface, but its boundary is irregular.

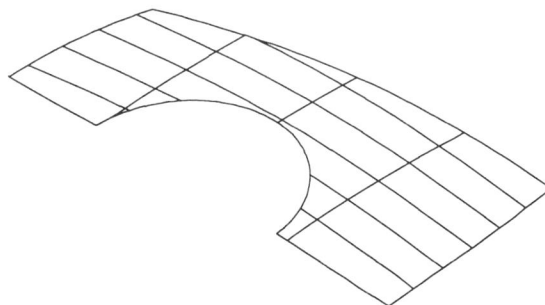

Figure 1.17 Surface model an automobile body panel

Given this problem, you might intuitively use the boundary curves that you can see as the defining wireframes to create the surface model. If you do that, you would probably get an irregular surface that resembles Figure 1.18.

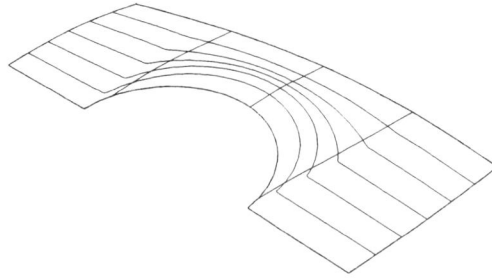

Figure 1.18 Irregular surface from irregular boundaries

Obviously, the surface shown in Figure 1.18 is not what we want as depicted in Figure 1.17. So, what has gone wrong? The answer is that a set of irregular curves will generate an irregular surface. You need to realize that unless the boundary lines are smooth wireframes, they cannot be used as defining curves of the surface.

To obtain a smooth surface with an irregular boundary, you have to do two steps. You need to use a set of smooth wireframes to produce a smooth surface that is much larger than the required surface, and then use the irregular boundary curve to trim the smooth surface.

In the computer memory, the resulting surface will consist of the original untrimmed smooth surface with smooth boundaries and the irregular boundaries. Although both of these data are saved in the electronic database, only the boundary and the remaining part of the trimmed surface are displayed. As a result, we obtain a smooth, free-form surface with irregular boundaries. This surface is called a trimmed surface.

In order to produce a free-form surface that is large enough for subsequent trimming, it is necessary to define a set of curves that encompass the required surface. To make such wireframes, you need to be able to visualize the existence of the defining curves that are outside the required surface.

In Figure 1.19, the creation of the smooth automobile body panel starts from a set of smooth wireframes. From the smooth wireframes, a smooth surface that is much larger than the required surface is made. To obtain the required surface, an irregular boundary is used to trim the large smooth surface.

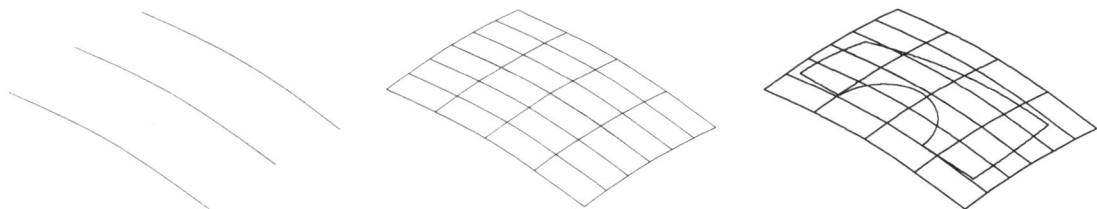

Figure 1.19 Defining curves (left), the untrimmed surface (center), and the irregular boundary (right)

To sum up, a smooth surface needs to be created from smooth defining wireframes. There are smooth surfaces with irregular edges in many designs. If you use the irregular edges to construct the surface directly, you will get a surface with many sudden changes in curvature. The surface will not be smooth at all.

To obtain a smooth surface with irregular edges, you should build a larger surface from smooth wireframes. Then, you trim the smooth surface with the irregular edges. The resulting surface is a trimmed surface.

In AutoSurf, you can trim a NURBS surface. A trimmed NURBS surface retains its original smooth defining boundaries in the database while possessing a new trimmed boundary. AutoSurf calls the original surface the base surface and the trimmed boundary the trim edges. The integrated data definition of the new trimmed surface contains both the base surface and the trim edges.

1.6   Enhanced Commands

To create and edit surfaces, you need 3D wireframes. To facilitate the creation of 3D wireframes, AutoSurf provides additional NURBS splines creation and editing commands that are not UCS dependent.

You can join and offset entities in the current view, and extend or trim them at the apparent intersection. The apparent intersection of two entities is the intersection that you see if you view them from a certain direction. Spatially, they are completely separated.

In AutoCAD, you can put entities into entity groups for ease of manipulation. With AutoSurf, you can control the visibility of individual entities without altering the display mode of the layers.

1.7   About this Book

This book guides you to use AutoSurf Release 3 as a design tool in NURBS surface model creation. Although there is a complete coverage of the AutoSurf commands, this book does not intend to replace the original AutoDesk training guides and manuals. The author presumes that you have acquired the AutoSurf application and installed the application properly in your computer.

There are six chapters in this book. The above delineation in this chapter is a brief introduction to surface modeling and the AutoSurf application. In Chapters 2 and 3, you will learn how to use AutoSurf to create the surface model of a video camera. In Chapter 2, you will practice using the wireframe creation and editing commands to make the wireframes for the surface model. In Chapter 3, you will apply AutoSurf NURBS surface creation and editing commands to put NURBS surfaces on the wireframes already created.

In order to enhance your learning, you will use a different approach in Chapter 4 to create a set of surfaces for a mobile phone. In addition, you will learn how to apply the utility commands on a surface model. In Chapter 5, you will inter-operate AutoCAD native solids with AutoSurf surfaces. You will also learn how to create an assembly of two surface models.

In a typical product development cycle, the electronic database of a surface model is required and is normally transmitted directly for subsequent operation. Documentation is not always a necessary step. However, you might still want to make an engineering document that consists of orthographic and isometric views. Therefore, you will learn how to prepare a document from a 3D surface model in Chapter 6.

Finally, the appendix provides a brief explanation of all the commands and variables provided by AutoSurf. You can use it as a quick reference.

1.8   Command Execution

After you have installed AutoSurf properly and used the default AutoSurf menu, you should have three additional pull-down menu items that are specific to AutoSurf — Surfaces, Drawings, and Mtools. If you do not see these pull-down menu items, use the MENULOAD command (Windows version) to add them to the Windows menu bar. See Figure 1.20.

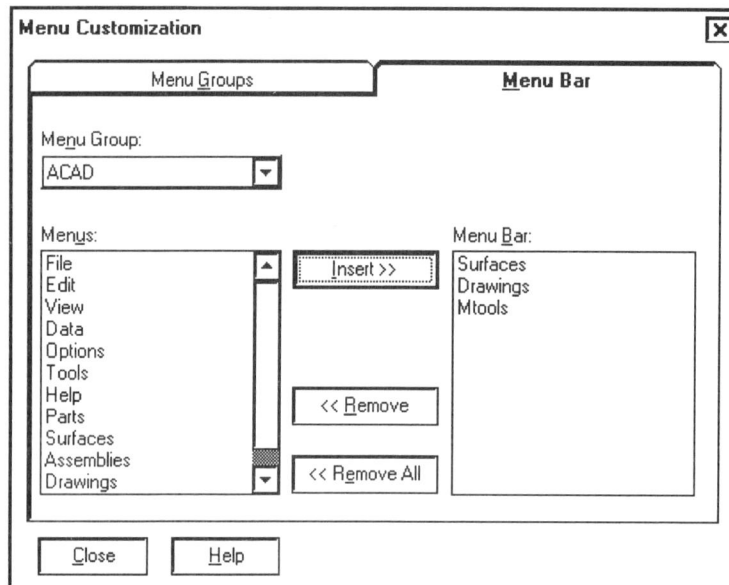

Figure 1.20 Loading AutoSurf pull-down menu items

Run the command. When the dialog box appears, select the Menu Bar page. Then, select the respective items for insertion.

Command: **MENULOAD**

In the Windows version, you will have four additional toolbars related to AutoSurf — Surface Create, Surface Edit, Drawing, and View. If you do not find these toolbars, select the Mtools pull-down menu, select the Mechanical Toolbars item, and then select each item there.

For example, suppose that you want to use the Surface Create toolbar. You should select the Mtools pull-down menu, select the Mechanical Toolbars item, and then select the Surface Create item. See Figure 1.21 and Figure 1.22.

\<Mtools\>   **\<Mechanical Toolbars\>**   **\<Surface Create\>**

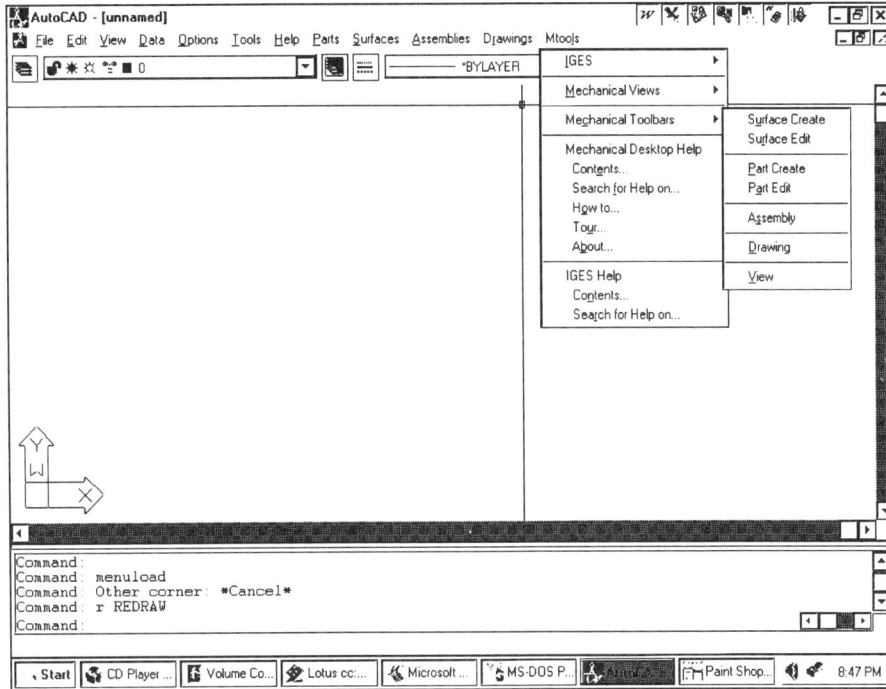

Figure 1.21 Bringing up the Surface Create toolbar

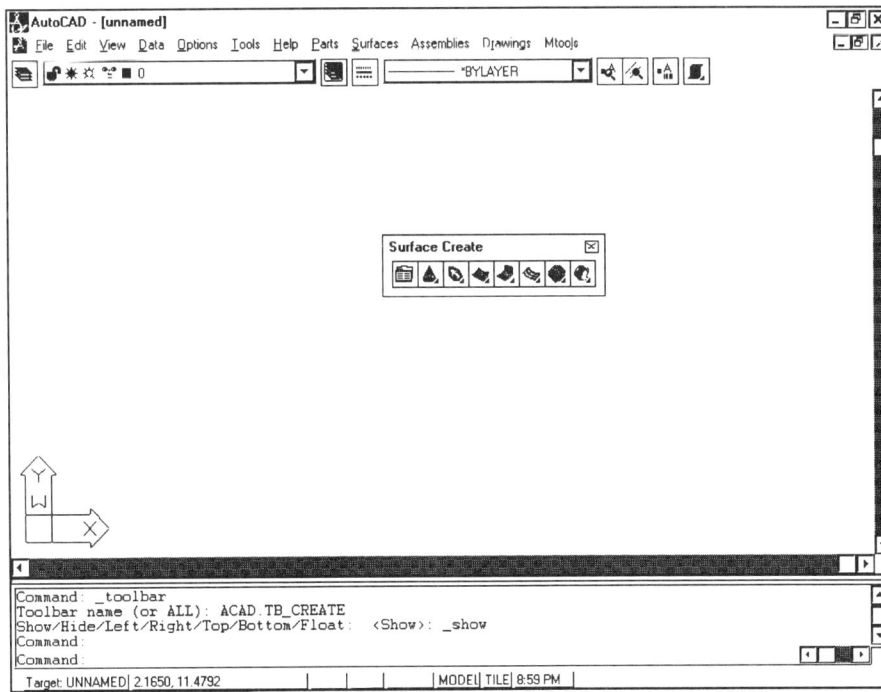

Figure 1.22 The Surface Create toolbar displayed

Note:
In the delineation that follows, <AAA> <BBB> will mean selecting the <AAA> pull-down menu, and then selecting the <BBB> item.

Command: **TOOLBAR**
Toolbar name (or ALL): ACAD.TB_CREATE
Show/Hide/Left/Right/Top/Bottom/Float: <Show>: _show

With the pull-down menu and toolbars in position, you can execute an AutoSurf command in several ways. You can use the pointing device to select an item from a pull-down menu or from a cascading menu in the pull-down menu. You can also use the pointing device to click an icon from the toolbars (Windows version). Finally, you can directly key in the command at the command window (Windows version) or the command prompt area (DOS version).

1.9 Registered Trademarks

The followings are registered trademarks of AutoDesk, Inc.:
AutoCAD
AutoCAD Designer
AutoSurf
AutoVision
3D Studio

Chapter 2

Wireframe Creation

A 3D NURBS surface model in a computer is a representation of a 3D object by a set of NURBS surfaces. To build the surface model of a given object, you should first analyze how many surfaces you need to create. You should consider using primitive surfaces, free-form surfaces, derived surfaces, or trimmed surfaces.

Except for primitive surfaces that you can begin right away, all other surfaces require that you prepare a set of wireframes to define the surface contour. Given a model to build, you should not expect the defining wireframes of the surfaces to be exhibited explicitly by the silhouettes or contours of the edges, especially in the case of trimmed surfaces.

As mentioned in the previous chapter, a trimmed surface is a smooth surface that usually possesses irregular edges. You need to start from a smooth surface using smooth boundaries, and then trim it with the required irregular edges. Thus, you must create all the defining edges although you can trim some of them away eventually.

This chapter, together with the next chapter, will guide you to create the surface model of a video camera. Figure 2.1 is a computer rendering of the completed model. In the proccss of making this model, you will gain an appreciation of how to build wireframes and surfaces with AutoCAD Release 13 together with AutoSurf Release 3 commands.

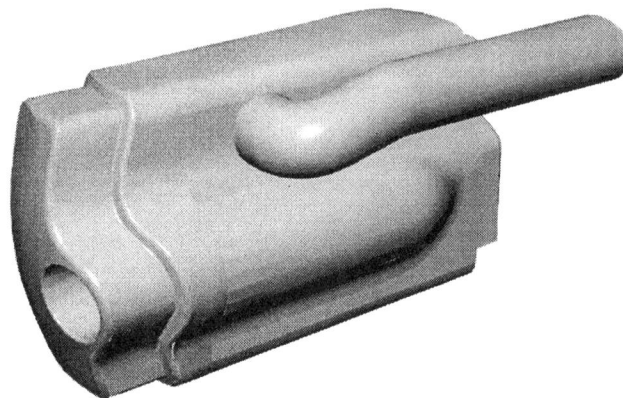

Figure 2.1 Video camera rendered in 3D Studio

You will perform wireframe creation for the model in this chapter, and surface creation in the next chapter. Figure 2.2 shows all the required wireframes for the model. If you compare Figure 2.1 with Figure 2.2 critically, you will notice that some of the defining wireframes in Figure 2.2 do not seem to appear on the surfaces of Figure 2.1. These wireframes are implicit. More practice is needed in order to gain experience in visualizing such implicit wireframes.

To work systematically, we separate the 3D model into four major parts: the frontal part, the rear part, the body part, and the eyepiece.

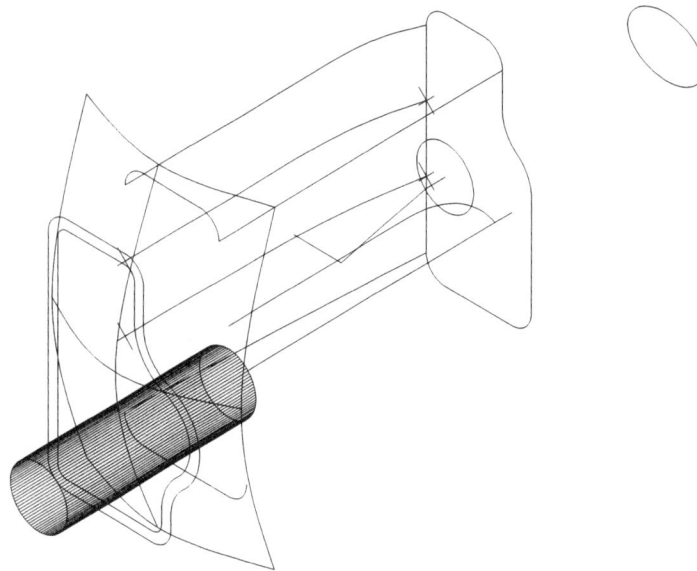

Figure 2.2 Completed wireframes for the video camera

This book is based on AutoSurf Release 3. You should first run the AMVER command to check what release number you are using. This information will also be needed for checking the compatibility of the 3D NURBS surface model you are going to create with other applications.

```
Command: AMVER
AutoSurf R3
```

As a good working habit, you should create separate layers for holding different objects in a drawing. Create a layer called WIRE1 with color cyan, and another layer called WIRE2 with color red. Set layer WIRE1 as the current layer.

Important Notes:
AutoSurf creates layers for storing entities used by the system. By default, the system freezes these layers. Do not thaw these layers or attempt to edit the content of them. Doing so will cause permanent corruption to your file.

<Data>   **<Layers...>**

Command: **DDLMODES**

Layer	Color
WIRE1	**cyan**
WIRE2	**red**

Current layer: **WIRE1**

2.1   Frontal Part of the Camera

Figure 2.3 shows the wireframes for the frontal part of the video camera that you are going to create. You can compare this with Figure 3.2 –– the surface model for the frontal part.

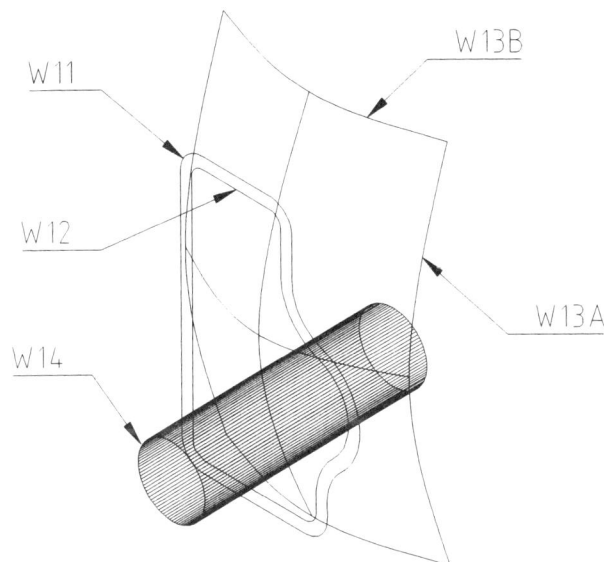

Figure 2.3 Wireframes for the frontal part of the video camera

You will create five groups of entities, together with two more groups (not shown in Figure 2.3) for the other parts of the camera. In the next chapter, you will use W13A and W13B to produce a smooth lofted UV surface, and from it create an offset surface. You will use W11, and W12 as wireframes to do several tasks –– projection, trimming, and extrusion. W14 is not a surface; it is a circle with thickness. You will use it to appreciate how to change an AutoCAD entity to an AutoSurf surface.

When you are working in 3D space in a computer, you can easily become lost if you do not know the orientation of the origin and the direction of the three axes. The UCS (User Coordinate System) icon is a very good tool for this situation. Turn on the UCS icon with the UCSICON command. Set the display of this icon at the origin position.

<Options>   **<UCS>**   **<Icon>**

Command: **UCSICON**
ON/OFF/All/Noorigin/ORigin <ON>: **ON**

<Options>   **<UCS>**   **<Icon Origin>**

Command: **UCSICON**
ON/OFF/All/Noorigin/ORigin <ON>: **OR**

Some AutoCAD commands provide the option of command line prompt or dialog box interaction. If you prefer the dialog box display, make sure that the system variable CMDDIA is set to 1.

Command: **CMDDIA**
New value for CMDDIA <1>: **1**

The operations of some AutoSurf commands depend on the setting of AutoSurf variables. Always run the AMSURFVARS command to set the system variables before starting to build a model.

<Surfaces>   **<Preferences...>**

Command: **AMSURFVARS**

If you do not know how to set the values of the variables, simply click the [Model Size] button and specify the size of the surface model. Watch the automatic setting of values. Click the [OK] button. See Figure 2.4.

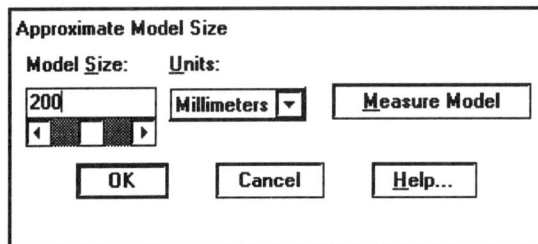

Figure 2.4 Setting the surface model size

Because you will set the entities on the XZ plane of the WCS, use the UCS command to change the orientation of the User Coordinate System. After that, set the display to the XZ plane of the WCS with the accelerator key [9].

[UCS]   **[X Axis Rotate UCS]**

Command: **UCS**
Origin/ZAxis/3point/OBject/View/X/Y/Z/Prev/Restore/Save/Del/?/<World>: **X**
Rotation angle about X axis <0>: **90**

Command: **9**

Other accelerator keys related to AutoSurf are:

F   Fits all the objects to the screen.
S   Starts the SPLINE command.
W   Toggles between drawing and model modes with the AMMODE command.
QQ   Edits the drawing views with the AMEDITVIEW command.
UU   Sets the UCS to the view with the UCS command.
VV   Hides or shows objects with the AMVISIBLE command.
1   Sets the display to a single viewport.
2   Sets the display to two viewports.
3   Sets the display to three viewports.
4   Sets the display to four viewports.
5   Sets the display to the top view.
6   Sets the display to the front view.
7   Sets the display to the right side view.
8   Sets the display to the isometric view.
9   Sets the display to the current UCS.
0   Performs the HIDE command.
[Rotates the view to the left.
]   Rotates the view to the right.
=   Rotates the view upward.
-   Rotates the view downward.

To begin with, run the LINE and CIRCLE commands to draw a series of line segments and a circle on the new UCS.

[Draw]   **[Line]**

Command: **LINE**
From point: **0,0**
To point: **@120<90**
To point: **@50<0**
To point: **@70<270**
To point: **[Enter]**

[Draw]   **[Line]**

Command: **LINE**
From point: **0,0**
To point: **@65<0**
To point: **@50<90**
To point: **[Enter]**

[Draw]   **[Circle Center Radius]**

Command: **CIRCLE**
3P/2P/TTR/<Center point>: **55,50**

Diameter/<Radius>: **25**

Look at your screen. If the current view window does not display all the entities created, use the accelerator key [F] to fit all the entities in the current view. See Figure 2.5.

Command: **F**

In the text that follows, we will not mention this command again. However, you should, from time to time, select this command or the ZOOM and PAN commands to ensure that all the entities appear clearly on the screen.

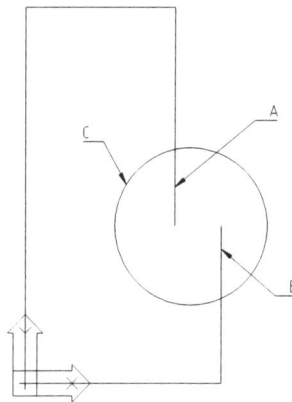

Figure 2.5 Lines and circle drawn

You need to trim the circle and two vertical lines. Use the TRIM command to do this. See Figure 2.6.

[Modify]   **[Trim]**

Command: **TRIM**
Select cutting edges: (Projmode = View, Edgemode = No extend)
Select objects: **[Select a point near (90,80).]**
Other corner: **[Select a point near (20,20).]**
Select objects: **[Enter]**
<Select object to trim>/Project/Edge/Undo: **[Select A (Figure 2.5).]**
<Select object to trim>/Project/Edge/Undo: **[Select B (Figure 2.5).]**
<Select object to trim>/Project/Edge/Undo: **[Select C (Figure 2.5).]**
<Select object to trim>/Project/Edge/Undo: **[Enter]**

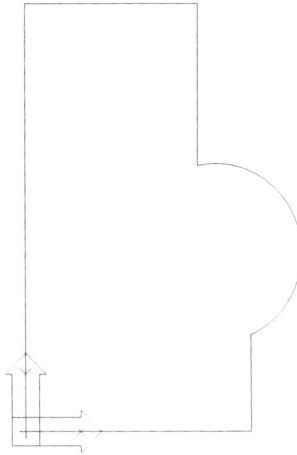

Figure 2.6 Circle and lines trimmed

In order to have a better view of the 3D wireframes that you are going to create, use the accelerator key [8] to set the display to an isometric view.

Command: **8**

As you continue to create more wireframe entities, selection of individual entities might become difficult. Naturally, you can put them into different layers. However, you might end up with a lot of layers, which can lead to confusion.

To solve the problem of entity manipulation, you can organize the entities into groups, and control their visibility individually rather than changing the visibility setting of the layer on which they reside.

Run the GROUP command to put the entities you created so far in an entity group named W1 for subsequent manipulation.

[**Standard Toolbar**]   [**Object Group**]

Command: **GROUP**

    [Group name: **W1**
    Selectable: **Yes**
    New:]

Select objects for grouping:
Select objects: [**Select all the entities.**]
Select objects: [**Enter**]

    [**OK**]

Use the COPY command to copy the entity group W1 to a distance of 40 units in the positive Z direction. See Figure 2.7.

[**Modify**]   [**Copy Object**]

```
Command: COPY
Select objects: GROUP
Enter group name: W1
Select objects: [Enter]
<Base point or displacement>/Multiple: 0,0,40
Second point of displacement: [Enter]
```

While you copy, you can specify the group name, as delineated above, or pick one members of a group to select the entire group, because the group is, by default, selectable. If you do not want the group anymore, you can use the Explode option. If you want to keep the group intact, but want to select a member of the group individually, rather than the entire group collectively, you can set the group to be unselectable.

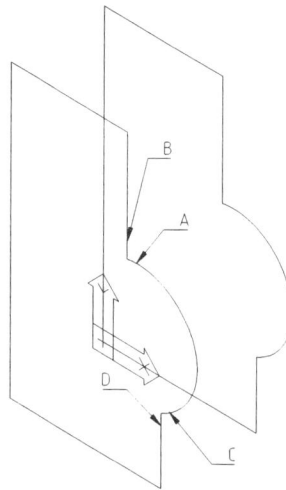

Figure 2.7 Entity group copied in the Z direction

As mentioned earlier, you can control the visibility of an individual entity. For the time being, the entity group W1 is not needed. It will be used later in the rear part of the camera. To hide it from view, you can use the AMVISIBLE command. After hiding, the group W1 becomes invisible.

<Surfaces>   **<Object Visibility...>**

Command: **AMVISIBLE**

      [Hide
       Select   **]**

Select objects to hide: **GROUP**
Enter group name: **W1**
Select objects to hide: **[Enter]**

      [OK]

When you copy grouped entities, the copied entities form an unnamed group. To edit these entities, you have to make the group unselectable or to explode the group. To explode a group means to ungroup the entities.

Run the GROUP command. Include the unnamed group in the listing, and then explode the unnamed group.

[Standard Toolbar]   [Object Group]

Command: **GROUP**

　　　　[Included Unnamed.
　　　　　Select the unnamed group with the prefix *.
　　　　　Explode
　　　　　OK　　　　　　　　　　　　　　　]

Edit the newly copied entities by running the AMFILLET3D command. This command is suitable for filleting a 3D wireframe because the current UCS setting does not affect this command.

<Surfaces>   <Edit Wireframe>   <Fillet>

Command: **AMFILLET3D**
(Radius = 0, Trim = Both)
Radius/Trim/<Select first object>: **R**
Radius: **20**
(Radius = 20, Trim = Both)
Radius/Trim/<Select first object>: **[Select A (Figure 2.7).]**
Radius/Trim/<Select second object>: **[Select B (Figure 2.7).]**

<Surfaces>   <Edit Wireframe>   <Fillet>

Command: **AMFILLET3D**
(Radius = 20, Trim = Both)
Radius/Trim/<Select first object: **[Select C (Figure 2.7).]**
Radius/Trim/<Select second object>: **[Select D (Figure 2.7).]**

After filleting two corners of the wireframe to R20 radius, repeat the AMFILLET3D command on the four remaining corners. This time, set the radius of fillet to 10 units. The completed wireframes should look like Figure 2.8.

<Surfaces>   <Edit Wireframe>   <Fillet>

Command: **AMFILLET3D**
(Radius = 0, Trim = Both)
Radius/Trim/<Select first object>: **R**
Radius: **10**
(Radius = 10, Trim = Both)
Radius/Trim/<Select first object>: **[Select the upper horizontal line.]**
Radius/Trim/<Select second object>: **[Select the left vertical line.]**

Repeat the AMFILLET3D command three more times.

<Surfaces> <Edit Wireframe> <Fillet>

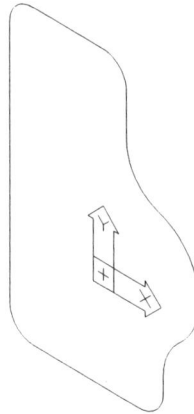

Figure 2.8 Corners filleted

After filleting, make a copy of the entities that exhibit on the screen with the COPY command. Copy them to a distance of -40 units in the Z direction. See Figure 2.9.

[**Modify**]   [**Copy Object**]

Command: **COPY**
Select objects: **[Select a point near (90,140).]**
Other corner: **[Select a point near (-60,-80).]**
Select objects: **[Enter]**
<Base point or displacement>/Multiple: **0,0,-40**
Second point of displacement: **[Enter]**

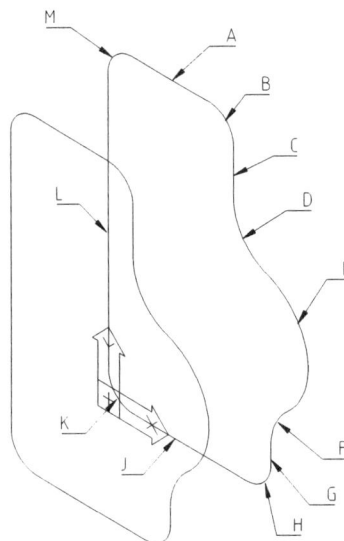

Figure 2.9 Entities copied

Put the newly copied entities in an entity group called W3 by using the GROUP command.

[Standard Toolbar] **[Object Group]**

Command: **GROUP**

    [Group name: **W3**
    Selectable: **Yes**
    New:]

Select objects for grouping:
Select objects: **[Select A through M (Figure 2.9).]**
Select objects: **[Enter]**

    [OK]

The entity group W3 will be used to create the wireframes for the body of the camera. Change its layer property to WIRE2.

<Edit> **<Properties...>**

Select objects: **GROUP**
Enter group name: **W3**
Select objects: **[Enter]**

[Properties
Layer...   **WIRE2**
OK   ]

The color of this entity group, after changing its layer property, should now be red, the color of layer WIRE2. Use the AMVISIBLE command to hide this group. We will come back to it later.

<Surfaces> **<Object Visibility...>**

Command: **AMVISIBLE**

    [**Hide**
    Select   ]

Select objects to hide: **GROUP**
Enter group name: **W3**
Select objects to hide: **[Enter]**

    [OK]

You are going to join the lines and arcs that remain on the screen to become a 3D polyline. Before you do so, take a look at the direction of each entity with the AMDIRECTION command.

<Surfaces>   <Edit Wireframe>   <Direction>

Command: **AMDIRECTION**
Select objects: **[Select all the entities on the screen.]**
Select objects: **[Enter]**
Reverse? Yes/<No>: **NO**

From the small arrowheads shown, you can see that the directions of the entities are not consistent. If you join them together manually, the result will be unpredictable. You have to either apply the DIRECTION command on some of them to reverse their direction, or use the AUTOMATIC option of the AMJOIN3D command during joining.

Refresh the screen with the REDRAW command.

[Standard Toolbar]   [Redraw View]

Command: **REDRAW**

Use the AMJOIN3D command to join the lines and arcs together to form a 3D polyline. This command can join entities together regardless of the current UCS setting. While running the command, use the Automatic mode to align the direction of all the input wireframes.

<Surfaces>   <Edit Wireframe>   <Join...>

Command: **AMJOIN3D**
       [Mode: **Automatic**
       Output: **3D Polyline**
       OK   ]

Select start wire: **[Select the lower horizontal line.]**
Select wires to join: **[Select all other entities.]**
Select wires to join: **[Enter]**

After joining and before the command completes, a small arrowhead will appear on the screen to depict the direction of the joined entity. Type No to accept.

Reverse? Yes/<No>: **NO**

Again, for the sake of ease in entity manipulation, use the GROUP command to put this polyline in an entity group called W11. This entity group will be used to trim a surface and be extruded to become an extruded surface.

[Standard Toolbar]   [Object Group]

Command: **GROUP**

      [Group name: **W11**
      Selectable: **Yes**
      New:]

Select objects for grouping:

Select objects: **LAST**
Select objects: **[Enter]**

   [OK]

In the last AMJOIN3D command, you have used the 3D Polyline option to obtain a polyline. If you want a spline instead, you can use the Spline option while joining, or you can use this polyline to fit to a spline with the AMFITSPLINE command as shown in Figure 2.10.

   <Surfaces>   **<Edit Wireframe>**   **<Spline Fit...>**

Command: **AMFITSPLINE**
Select wires: **GROUP**
Enter group name: **W11**
Select wires: **[Enter]**

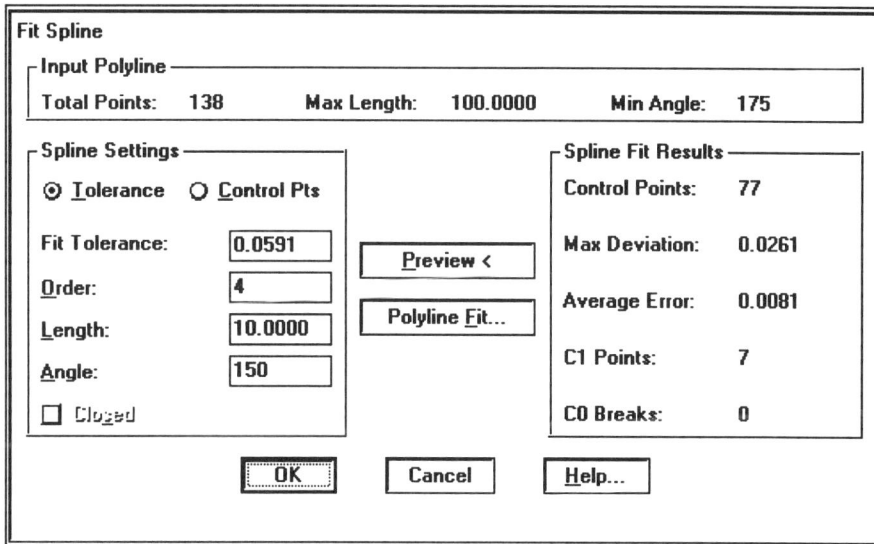

Figure 2.10 AMFITSPLINE dialog box

[Spline Settings:
   Length: **10**
   Preview   ]

Enter RETURN to continue: **[Enter]**

   [OK]

A polyline is an entity with lines and arcs joined together at their ends. To fit a polyline to become a spline is not a simple conversion of data type, because even the tangential line/arc or arc/arc segments of a polyline are not as continuously smooth as a spline. In terms of smoothness or continuity, there are three classes: C0, C1, and C2.

C0 class consists of a sudden change in curvature.

C1 class consists of changes in curvature at tangential points.

C2 class is continuously smooth.

Naturally, a polyline could be C0 only if it consists of nontangential line/arc or arc/arc segments, or C1 only if all of its line/arc and arc/arc joins are tangential. Unfortunately, it does not fall in the C2 class. Unlike a polyline, a spline belongs to the C2 class.

Consequently, fitting a polyline to a spline can be only approximate because data from the input polyline is extracted to construct a spline. The question is how.

When you use the AMFITSPLINE command to fit a polyline to become a spline, it displays the Total Points (number of points), the Max Length (length of the longest segment), and the Min Angle (sharpest angle in the polyline).

You can either set the tolerance value or the number of control points for spline fitting. The tolerance value relates to the maximum deviation of the spline to the original polyline. A closer tolerance creates a spline that more closely resembles the shape of the original polyline. The system variable AMPFITTOL sets its default value. The number of control points is the number of control points of the resulting spline.

The order of the spline is the degree of the highest exponent plus one in its mathematical equation. Order 4 gives the smoothest result.

The Length of fit and the Angle of fit set the system variables AMPFITLEN and AMPFITANG respectively. When a segment of the input polyline is shorter than the specified fit length, that segment will remain flat in the spline. When the angle between two segments of the input polyline is smaller than the specified fit angle, that corner breaks the resulting spline into two.

If you click the [Closed] button, the resulting spline will be a closed spline even though the input polyline is open ended.

Naturally, a surface belongs to the same category of the curves that creates it. For example, a C0 curve will produce a C0 surface. You must note that not all applications support entities with C0 or C1 continuity. You should check compatibility beforehand if you export C0 or C1 entities to other applications.

If you use C0 curves for the generation of a surface, AutoSurf will usually break it into separate components at the abrupt changes of curvature. As a result, AutoSurf produces a series of C1 or C2 surfaces instead of a single C0 surface. When this happens, it will display the following warning message:

Warning:
?? surfaces will be created from the input wires.

If you use the Spline option in the AMJOIN3D command, the entities will be first joined together, and then fitted to a spline in much the same way as the AMFITSPLINE command does.

Contrary to fitting a polyline to a spline, you can revert a spline to a polyline. Use the AMUNSPLINE command on the spline to change it back to a polyline. Because we need a spline here, select the U command to undo this operation.

<Surfaces>   **<Edit Wireframe>**   **<Unspline>**

Command: **AMUNSPLINE**

[Standard Toolbar]   **[Undo]**
Command: **U**

Run the LIST command to confirm that the entity is a spline before proceeding.

[Inquiry]   **[List]**

Command: **LIST**
Select objects: **LAST**
Select objects: **[Enter]**

      SPLINE Layer: WIRE1
        Space: Model space
      Handle = 13B
      Group = W11
      Circumference: 355.1101
        Order: 4
       Properties: Planar, Non-Rational, Non-Periodic
      Parametric Range: Start 0.0000
          End 355.1339
      Control Points: X = 10.0000 , Y = 0.0000 , Z = 40.0000
         X = 25.0000 , Y = 0.0000 , Z = 40.0000
         X = 40.0000 , Y = 0.0000 , Z = 40.0000
        ....

Next, you will create an offset of the spline. Switch to the front view using the accelerator key [6]. Run the OFFSET command to create a new spline at 5 units offset inside the existing one. See Figure 2.11.

Command: **6**

[Modify]   **[Offset]**

Command: **OFFSET**
Offset distance or Through <Through>: **5**
Select object to offset: **[Select the spline.]**
Side to offset? **[Select a point inside the spline.]**
Select object to offset: **[Enter]**

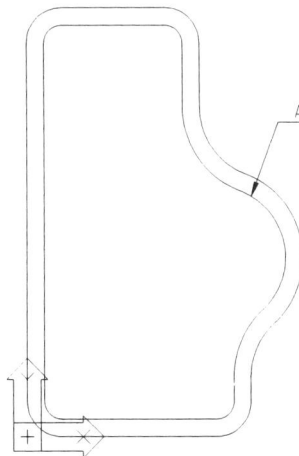

Figure 2.11 Spline offset

After doing the offset, put the new spline in an entity group called W12. This entity group will be used to project a wireframe on a surface.

[Standard Toolbar]   **[Object Group]**

Command: **GROUP**

    [Group name: **W12**
    Selectable: **Yes**
    New:]

Select objects for grouping:
Select objects: **[Select A (Figure 2.11).]**
Select objects: **[Enter]**

    [OK]

Return to an isometric view.

Command: **8**

Hide the two splines with the AMVISIBLE command. After hiding, your screen should be blank.

<Surfaces>   **<Object Visibility...>**

Command: **AMVISIBLE**

    [Hide
     Select   ]

Select objects to hide: **GROUP**
Enter group name: **W11,W12**
Select objects to hide: **[Enter]**

    [OK]

In addition to using the AMFITSPLINE command to fit a 3D polyline to become a spline, you can create a spline directly with the SPLINE command. Create six splines. See Figure 2.12.

[Polyline]   **[Spline]**

Command: **SPLINE**
Object/<Enter first point>: **-20,-20**
Enter point: **-20,60,18**
Close/Fit Tolerance/<Enter point>: **-20,140**
Close/Fit Tolerance/<Enter point>: **[Enter]**
Enter start tangent: **[Enter]**
Enter end tangent: **[Enter]**

[Modify]   **[Copy Object]**

Command: **COPY**
Select objects: **LAST**
Select objects: **[Enter]**
<Base point or displacement>/Multiple: **100,0**
Second point of displacement: **[Enter]**

[Polyline]   [Spline]

Command: **SPLINE**
Object/<Enter first point>: **30,-20,12**
Enter point: **30,60,35**
Close/Fit Tolerance/<Enter point>: **30,140,12**
Close/Fit Tolerance/<Enter point>: **[Enter]**
Enter start tangent: **[Enter]**
Enter end tangent: **[Enter]**

[Polyline]   [Spline]

Command: **SPLINE**
Object/<Enter first point>: **MID** of **[Select the left spline (Figure 2.12).]**
Enter point: **MID** of **[Select the middle spline (Figure 2.12).]**
Close/Fit Tolerance/<Enter point>: **MID** of **[Select the right spline (Figure 2.12).]**
Close/Fit Tolerance/<Enter point>: **[Enter]**
Enter start tangent: **[Enter]**
Enter end tangent: **[Enter]**

[Polyline]   [Spline]

Command: **SPLINE**
Object/<Enter first point>: **END** of **[Select the lower end of the left spline (Figure 2.12).]**
Enter point: **END** of **[Select the lower end of the middle spline (Figure 2.12).]**
Close/Fit Tolerance/<Enter point>: **END** of **[Select the lower end of the right spline (Figure 2.12).]**
Close/Fit Tolerance/<Enter point>: **[Enter]**
Enter start tangent: **[Enter]**
Enter end tangent: **[Enter]**

[Modify]   **[Copy Object]**

Command: **COPY**
Select objects: **LAST**
Select objects: **[Enter]**
<Base point or displacement>/Multiple: **0,160**
Second point of displacement: **[Enter]**

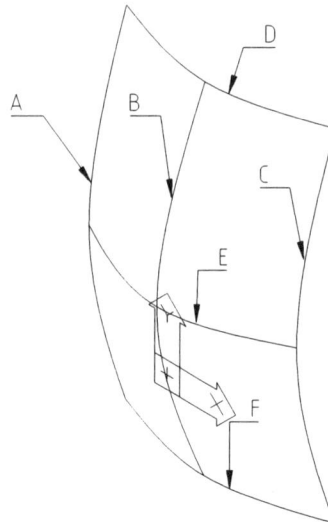

Figure 2.12 Six splines created

Put the three vertical splines and the three horizontal splines into two entity groups, called W13A and W13B respectively, by using the GROUP command. These two entity groups will be used to create a lofted UV surface.

[Standard Toolbar]   **[Object Group]**

Command: **GROUP**

      [Group name: **W13A**
      Selectable: **Yes**
      New:]

Select objects for grouping:
Select objects: **[Select A (Figure 2.12).]**
Select objects: **[Select B (Figure 2.12).]**
Select objects: **[Select C (Figure 2.12).]**
Select objects: **[Enter]**

      [Group name: **W13B**
      Selectable: **Yes**
      New:]
Select objects for grouping:
Select objects: **[Select D (Figure 2.12).]**
Select objects: **[Select E (Figure 2.12).]**
Select objects: **[Select F (Figure 2.12).]**
Select objects: **[Enter]**

    [OK]

To complete the frontal part, create a circle with the CIRCLE command, and change its thickness property to 50 units. See Figure 2.13.

[Draw]   [Circle Center Radius]

Command: **CIRCLE**
3P/2P/TTR/<Center point>: **55,50**
Diameter/<Radius>: **15**

<Edit>   <Properties...>

Select objects: **LAST**
Select objects: **[Enter]**

[Properties
Thickness   **100**
OK   ]

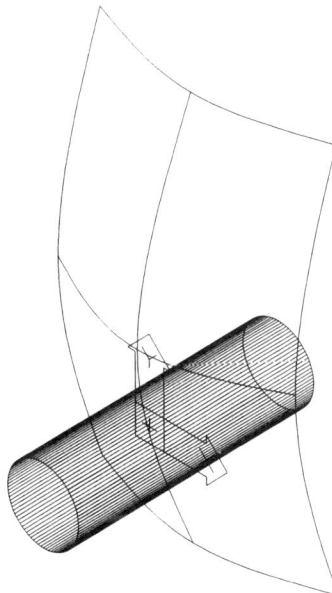

Figure 2.13 Circle with thickness property of 50 units

Apply the GROUP command to assign this circle to an entity group called W14. This entity group will be used to convert an AutoCAD entity to an AutoSurf surface.

[Standard Toolbar]   [Object Group]

Command: **GROUP**

        [Group name: **W14**
         Selectable: **Yes**
         New:]

Select objects for grouping:
Select objects: **LAST**
Select objects: **[Enter]**

        [OK]

You have completed the wireframes for the frontal part. You have created seven sets of entities, and have put them in seven entity groups: W1, W11, W12, W13A, W13B, W14, and W3. With the exception of groups W1 and W3, these groups will be used to make the surface model of the frontal part of the camera.

To sum up, W11 will be used to trim a surface and be extruded to form an extruded surface, W12 will be used to project a wireframe on a surface, W13A and W13B will be used to create a lofted UV surface, and W14 will be used to convert an AutoCAD entity to an AutoSurf surface.

To proceed, apply the AMVISIBLE command to hide the entity groups W13A, W13B, and W14. Then unhide W1. See Figure 2.14.

<Surfaces>   <Object Visibility...>

Command: **AMVISIBLE**

      [Hide
       Select   ]

Select objects to hide: **GROUP**
Enter group name: **W13A,W13B,W14**
Select objects to hide: **[Enter]**

      [Unhide
       Select   ]

Select objects to unhide: **GROUP**
Enter group name: **W1**
Select objects to unhide: **[Enter]**

      [OK]

You will modify the entities in the entity group W1 to become the wireframes for the rear part of the camera.

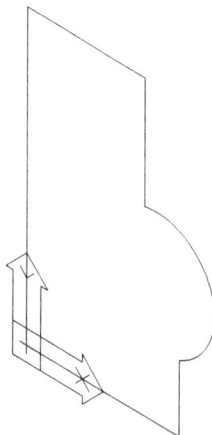

Figure 2.14 Entity groups W13A, W13B and W14 hidden, and W1 unhidden

2.2   **Rear Part of the Camera**

Figure 2.15 shows the completed wireframes for the rear part of the video camera. This part of the camera consists of a flat planar surface with a box-like object attached to it. The box-like object does not require any wireframes. You need only the simple planar wireframe that is modified from the entity group W1.

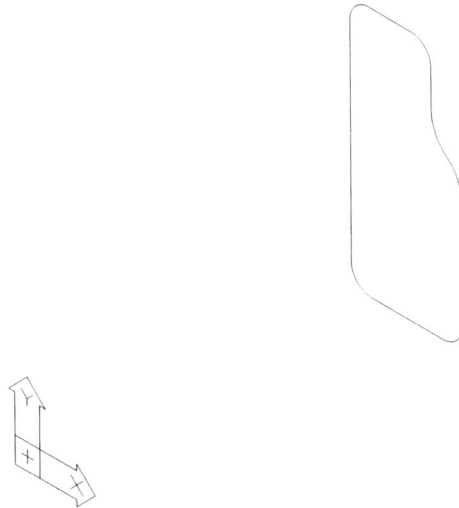

Figure 2.15 Wireframes for the rear part of the video camera

You will continue to work on layer WIRE1. On the screen, you have the entity group W1. Use the MOVE command to move the group of entities a distance of 150 units in the negative Z direction. See Figure 2.16. After moving the entities, apply the GROUP command with the Explode option to ungroup W1.

```
[Modify]                    [Move]
Command: MOVE
Select objects: GROUP
Enter group name: W1
Select objects: [Enter]
Base point or displacement: 0,0,-150
Second point of displacement: [Enter]


[Standard Toolbar]     [Object Group]


Command: GROUP

        [Group name:        W1
         Explode
         OK                        ]
```

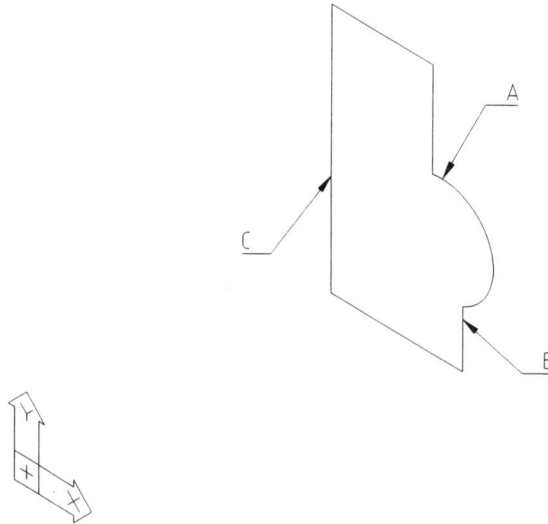

Figure 2.16 Group W1 moved and ungrouped

Use the EXTEND command to extend the right vertical line to meet the arc. Then, apply the MOVE command to move the left vertical line a distance of 10 units to the right. See Figure 2.17.

[Modify]   **[Extend]**

Command: **EXTEND**
Select boundary edges: (Projmode = View, Edgemode = No extend)
Select objects: **[Select A (Figure 2.16).]**
Select objects: **[Enter]**
<Select object to extend>/Project/Edge/Undo: **[Select B (Figure 2.16).]**
<Select object to extend>/Project/Edge/Undo: **[Enter]**

[Modify]   [Move]

Command: **MOVE**
Select objects: **[Select C (Figure 2.16).]**
Select objects: **[Enter]**
Base point or displacement: **10<0**
Second point of displacement: **[Enter]**

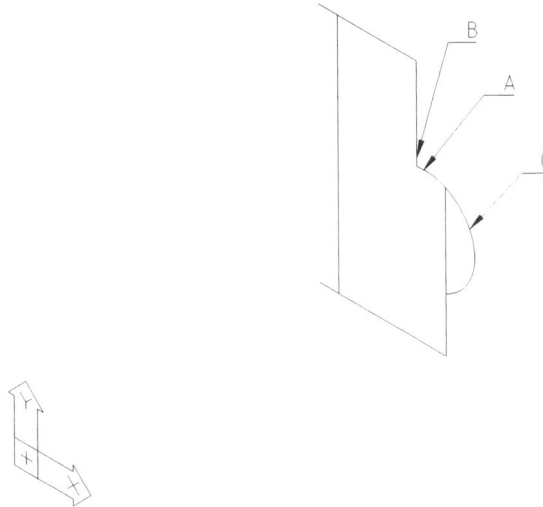

Figure 2.17 Lines extended and moved

Use the AMFILLET3D command to create a fillet of radius 20 units between the central vertical line and the upper portion of the arc. Select the ERASE command to erase the arc, and then select the AMFILLET3D command again to form a fillet between the new fillet and the right vertical line. See Figure 2.18.

<Surfaces>　　　**<Edit Wireframe>**　　　　**<Fillet>**

Command: **AMFILLET3D**
(Radius = 10, Trim = Both)
Radius/Trim/<Select first object>: **R**
Radius <10>: **20**
(Radius = 20, Trim = Both)
Radius/Trim/<Select first object>: **[Select A (Figure 2.17).]**
Radius/Trim/<Select second object>: **[Select B (Figure 2.17).]**

[Modify]　　　　　　**[Erase]**

Command: **ERASE**
Select objects: **[Select C (Figure 2.17).]**
Select objects: **[Enter]**

<Surfaces>　　　**<Edit Wireframe>**　　　　**<Fillet>**

Command: **AMFILLET3D**
(Radius = 20, Trim = Both)
Radius/Trim/<Select first object>: **[Select the lower end of the fillet arc.]**
Radius/Trim/<Select second object>: **[Select the right vertical line.]**

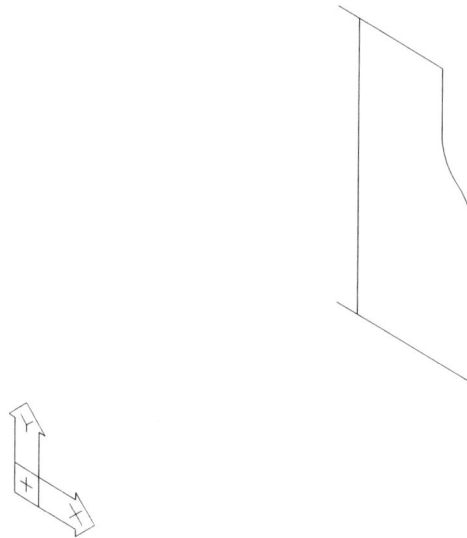

Figure 2.18 Two corners filleted and the arc erased

Repeat the AMFILLET3D command on the four remaining corners of the geometry with a radius of 10 units. See Figure 2.19.

<Surfaces>   <Edit Wireframe>   <Fillet>

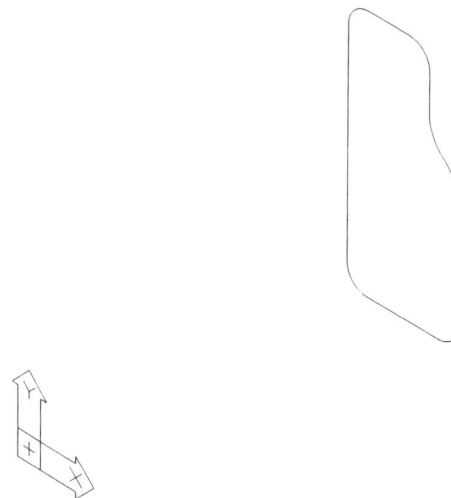

Figure 2.19 Completed polyline

These are all the entities necessary for the rear part of the video camera. Before proceeding to join them together, use the COPY command to copy these entities for the body of the camera, because they are common to the rear portion and the body.

[Modify]   [Copy Object]

Command: **COPY**

Select objects: **[Select all the entities on the screen.]**
Select objects: **[Enter]**
<Base point or displacement>/Multiple: **0,0**
Second point of displacement: **[Enter]**

Now, there are two sets of entities located at the same position. One set is for the rear part, and the other set is for the body of the camera. To separate them, change the original set of entities to layer WIRE2.

<Edit>   **<Properties...>**

Select objects: **P**
Select objects: **[Enter]**

[Properties
Layer...   **WIRE2**
OK   ]

Turn off the layer WIRE2. Now, you have only one set of entities left on the screen.

**[Object Properties
Layer Control
WIRE2**   **OFF**   ]

Refer to Figure 2.19. Join the entities together to become a 3D polyline by using the AMJOIN3D command.

<Surfaces>   **<Edit Wireframe>**   **<Join...>**

Command: **AMJOIN3D**

        [Mode: **Automatic**
        Output: **3D Polyline**
        OK   ]

Select start wire: **[Select the lower horizontal line (Figure 2.19).]**
Select wires to join: **[Select all other entities (Figure 2.19).]**
Select wires to join: **[Enter]**
Reverse? Yes/<No>: **NO**

Put this polyline in an entity group named W21 using the GROUP command. This entity group will be used to create a trimmed planar surface. Now, you have finished the wireframes for the rear part.

[Standard Toolbar]   **[Object Group]**

Command: **GROUP**

        [Group name: **W21**
        Selectable: **Yes**
        New:]

Select objects for grouping:

Select objects: **LAST**
Select objects: **[Enter]**

　　　[OK]

Hide the entity group W21 with the AMVISIBLE command.

<Surfaces> 　　**<Object Visibility...>**

Command: **AMVISIBLE**

　　**[Hide
　　Select** 　　　　　　　　]

Select objects to hide: **GROUP**
Enter group name: **W21**
Select objects to hide: **[Enter]**

　　　[OK]

　　You have created two sets of entities; one set is on layer WIRE2, and one set is in entity group W21. The set of entities on layer WIRE2 will be used to make the wireframes of the body of the camera. The entity group W21 will be used to make a trimmed planar surface for the rear part of the camera.

　　Because you have hidden the entity group W21, and have turned off layer WIRE2, there should be nothing visible on the screen.

2.3   Body of the Camera

Figure 2.20 shows the completed wireframes for the body of the video camera. The body consists of four sides. You will create four sets of wireframes.

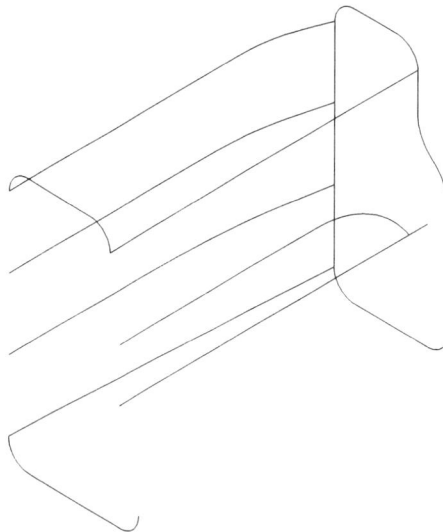

Figure 2.20 Wireframes for the body of the video camera

Turn on layer WIRE2 and set the current layer to it.

<Data>   **<Layers...>**

Command: **DDLMODES**

Layer	
WIRE2	**On**

Current layer: **WIRE2**

Use the AMVISIBLE command to unhide the entity group W3, and the GROUP command to explode the group. See Figure 2.21.

<Surfaces>   **<Object Visibility...>**

Command: **AMVISIBLE**

    [Unhide
     Select   **]**

Select objects to unhide: **GROUP**
Enter group name: **W3**
Select objects to unhide: **[Enter]**

    [OK]

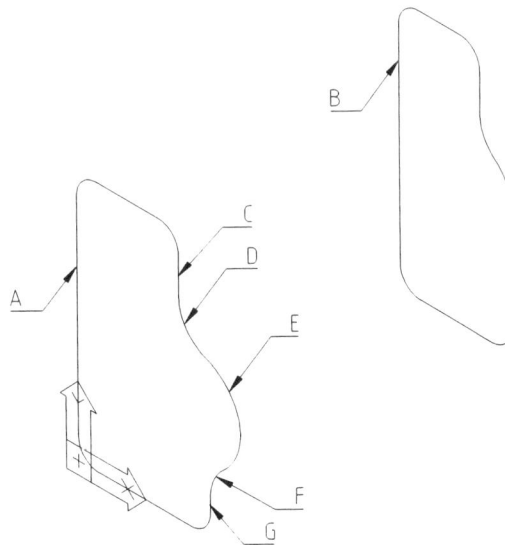

Figure 2.21 Layer WIRE2 turned on and group W3 unhidden

[Standard Toolbar]   **[Object Group]**

Command: **GROUP**

[Select the group W3.
Explode.
OK   ]

During drawing creation, you might need to check the data already drawn with the LIST and DIST command. AutoSurf has a command to check the distances between objects. Use the AMCHECKFIT command to find out the minimum distance between the left vertical lines of the two wireframes.

<Surfaces>   <Utilities>   <Check Fit>

Command: **AMCHECKFIT**
Select check wires: **[Select A (Figure 2.21).]**
Select check wires: **[Enter]**
Select target wire or surface: **[Select B (Figure 2.21).]**
Minimum distance = 150.333, Maximum distance = 150.333
(Scale = 0.1404)
Graph/Scale/Table/<eXit>: **[Enter]**

If you compare Figure 2.20 with Figure 2.21, you will find that some entities are not required. Run the ERASE command to remove the unwanted entities. See Figure 2.22.

[Modify]   [Erase]

Command: **ERASE**
Select objects: **[Select C, D, E, F, and G (Figure 2.21).]**
Select objects: **[Enter]**

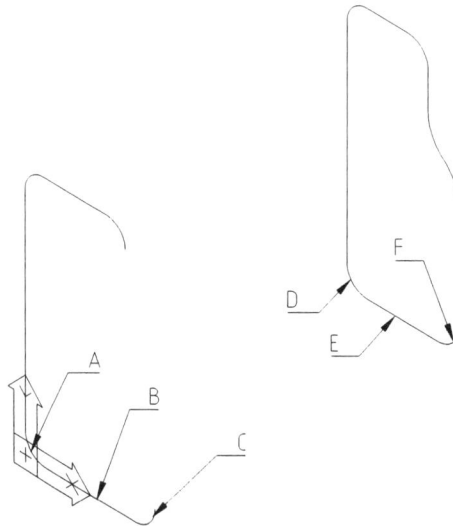

Figure 2.22 Some entities erased

Refer to Figure 2.22. Use the AMJOIN3D command to join six segments into two 3D polylines.

\<Surfaces\>   **\<Edit Wireframe\>**   **\<Join...\>**

Command: **AMJOIN3D**

       [Mode: **Automatic**
       Output: **3D Polyline**
       OK   ]

Select start wire: **[Select A (Figure 2.22).]**
Select wires to join: **[Select B (Figure 2.22).]**
Select wires to join: **[Select C (Figure 2.22).]**
Select wires to join: **[Enter]**
Reverse? Yes/\<No\>: **NO**

\<Surfaces\>   **\<Edit Wireframe\>**   **\<Join...\>**

Command: **AMJOIN3D**

       [Mode: **Automatic**
       Output: **3D Polyline**
       OK   ]

Select start wire: **[Select D (Figure 2.22).]**
Select wires to join: **[Select E (Figure 2.22).]**
Select wires to join: **[Select F (Figure 2.22).]**
Select wires to join: **[Enter]**
Reverse? Yes/\<No\>: **NO**

Put the two newly joined 3D polylines in an entity group named W31, using the GROUP command. This entity group will be used to make a ruled surface for the bottom of the body of the camera.

[Standard Toolbar]   **[Object Group]**

Command: **GROUP**

       [Group name: **W31**
       Selectable: **Yes**
       New:]

Select objects for grouping:
Select objects: **[Select A (Figure 2.22).]**
Select objects: **[Select D (Figure 2.22).]**
Select objects: **[Enter]**

       [OK]

\<Surfaces\>   **\<Object Visibility...\>**

Hide this entity group with the AMVISIBLE command. See Figure 2.23.

Command: **AMVISIBLE**

[**Hide**
   Select   ]

Select objects to hide: **GROUP**
Enter group name: **W31**
Select objects to hide: **[Enter]**

    [OK]

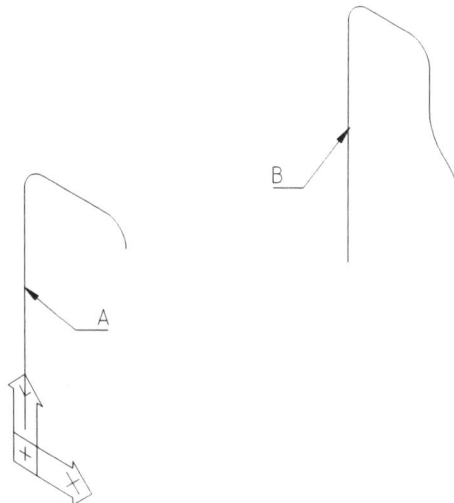

Figure 2.23 Two 3D polylines hidden

Change the display form of point entities by manipulating the variable PDMODE. Set its value to 3. Then, use the DIVIDE command to create two points along the two left vertical lines of the frontal and rear components. See Figure 2.24.

<Options>   **<Display>**   **<Point Style...>**

Command: **PDMODE**
New value for PDMODE: **3**

[Point]   **[Divide]**

Command: **DIVIDE**
Select object to divide: **[Select the line A (Figure 2.23).]**
<Number of segments>/Block: **3**

[Point]   **[Divide]**

Command: **DIVIDE**
Select object to divide: **[Select the line B (Figure 2.23).]**
<Number of segments>/Block: **3**

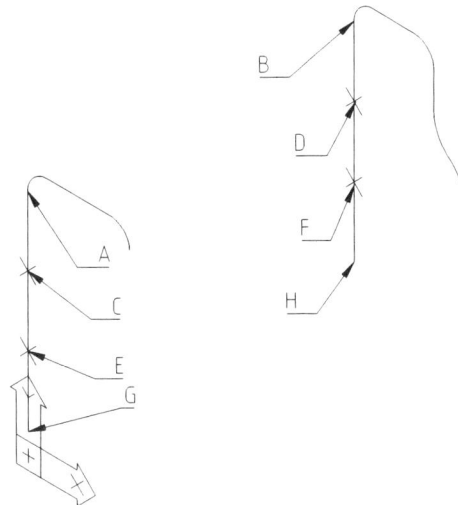

Figure 2.24 Points created

Use the SPLINE command to create three splines, and then use the LINE command to create a straight line segment. See Figure 2.25.

[**Polyline**]   [**Spline**]

Command: **SPLINE**
Object/<Enter first point>: **END** of **[Select A (Figure 2.24).]**
Enter point: **@1,0,-100**
Close/Fit Tolerance/<Enter point>: **@2,0,-30**
Close/Fit Tolerance/<Enter point>: **END** of **[Select B (Figure 2.24).]**
Close/Fit Tolerance/<Enter point>: **[Enter]**
Enter start tangent: **[Enter]**
Enter end tangent: **[Enter]**

[**Polyline**]   [**Spline**]

Command: **SPLINE**
Object/<Enter first point>: **NODE** of **[Select C (Figure 2.24).]**
Enter point: **@1,0,-80**
Close/Fit Tolerance/<Enter point>: **@2,0,-40**
Close/Fit Tolerance/<Enter point>: **NODE** of **[Select D (Figure 2.24).]**
Close/Fit Tolerance/<Enter point>: **[Enter]**
Enter start tangent: **[Enter]**
Enter end tangent: **[Enter]**

[**Polyline**]   [**Spline**]

Command: **SPLINE**
Object/<Enter first point>: **NODE** of **[Select E (Figure 2.24).]**
Enter point: **@1,0,-60**
Close/Fit Tolerance/<Enter point>: **@2,0,-45**
Close/Fit Tolerance/<Enter point>: **NODE** of **[Select F (Figure 2.24).]**
Close/Fit Tolerance/<Enter point>: **[Enter]**
Enter start tangent: **[Enter]**

Enter end tangent: **[Enter]**

[Draw]   [Line]

Command: **LINE**
From point: **END** of **[Select G (Figure 2.24).]**
To point: **END** of **[Select H (Figure 2.24).]**
To point: **[Enter]**

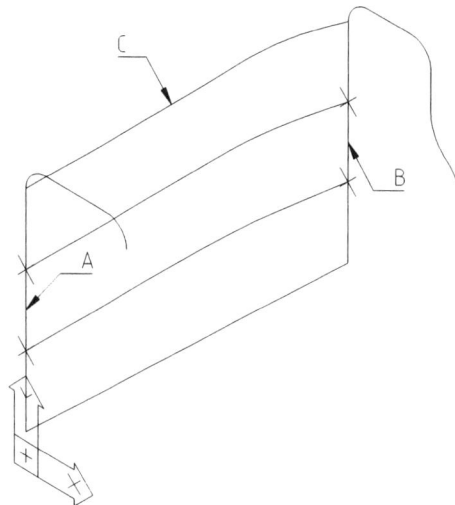

Figure 2.25 Three splines and a straight line created

The two vertical lines are not needed. Use the ERASE command to delete them. Then, make a copy of the uppermost spline with the COPY command. The copied spline belongs to the top surface of the camera body. See Figure 2.26.

[Modify]   **[Erase]**

Command: **ERASE**
Select objects: **[Select A (Figure 2.25).]**
Select objects: **[Select B (Figure 2.25).]**
Select objects: **[Enter]**

[Modify]   **[Copy Object]**

Command: **COPY**
Select objects: **[Select C (Figure 2.25).]**
Select objects: **[Enter]**
<Base point or displacement>/Multiple: **0,0**
Second point of displacement: **[Enter]**

To separate the newly copied spline, change it to layer WIRE1. Then, turn off layer WIRE1.

<Edit>   <Properties...>

Select objects: **LAST**
Select objects: **[Enter]**

[Properties
Layer...   **WIRE1**
OK   ]

[**Object Properties**
Layer Control
WIRE1   **OFF**   ]

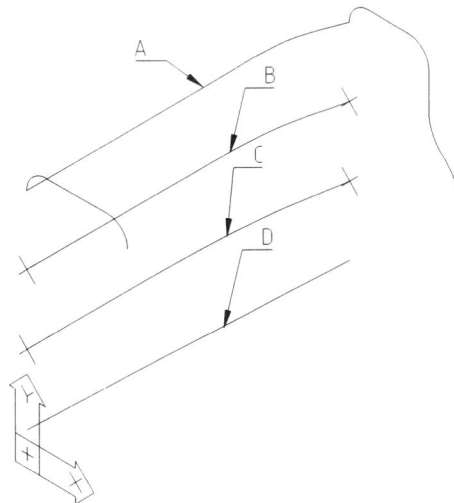

Figure 2.26 Entities for entity group W32

Refer to Figure 2.26, use the GROUP command to put three splines and one line in an entity group called W32. This entity group will be used to make a lofted U surface for the left side of the body of the camera.

    [Standard Toolbar]   **[Object Group]**

Command: **GROUP**

        [Group name: **W32**
         Selectable: **Yes**
         New:]

Select objects for grouping:
Select objects: **[Select A, B, C, and D (Figure 2.26).]**
Select objects: **[Enter]**

        [OK]

Use the AMVISIBLE command to hide the entity group W32. Then, turn on layer WIRE1 again. See Figure 2.27.

<Surfaces>   <Object Visibility...>

Command: **AMVISIBLE**

　　　[Hide
　　　　Select   ]

Select objects to hide: **GROUP**
Enter group name: **W32**
Select objects to hide: **[Enter]**

　　　[OK]

[Object Properties
Layer Control
WIRE1   ON   ]

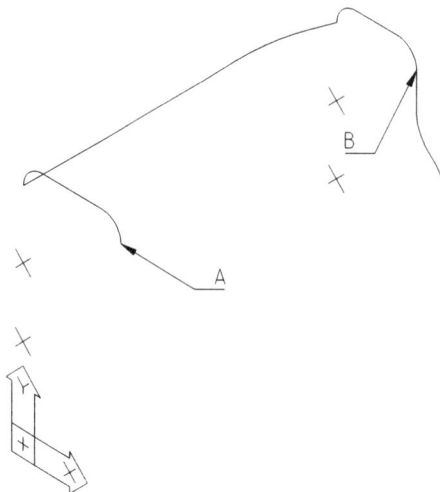

Figure 2.27 Entity group W32 hidden and layer WIRE1 turned on

Create a line with the LINE command. See Figure 2.28.

[Draw]   [Line]

Command: **LINE**
From point: **END** of **[Select A (Fig 2.27).]**
To point: **END** of **[Select B (Fig 2.27).]**
To point: **[Enter]**

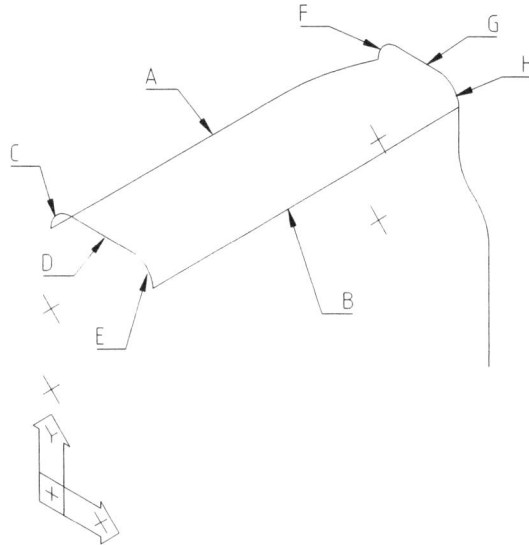

Figure 2.28 Line added

Put this line and the spline in an entity group W33A with the GROUP command.

    [Standard Toolbar]   **[Object Group]**

Command: **GROUP**

      [Group name: **W33A**
      Selectable: **Yes**
      New:]

Select objects for grouping:
Select objects: **[Select A and B (Figure 2.28).]**
Select objects: **[Enter]**

    [OK]

Refer to Figure 2.28. Use the AMJOIN3D command to join segments C, D, and E to become a 3D polyline and join segments F, G, and H to become another 3D polyline. Put the two 3D polylines in an entity group W33B using the GROUP command.

    <Surfaces>   **<Edit Wireframe>**   **<Join...>**

Command: **AMJOIN3D**

      [Mode: **Automatic**
      Output: **3D Polyline**
      OK   ]

Select start wire: **[Select C (Figure 2.28).]**
Select wires to join: **[Select D (Figure 2.28).]**
Select wires to join: **[Select E (Figure 2.28).]**
Select wires to join: **[Enter]**

Reverse? Yes/<No>: **NO**

<Surfaces>   <Edit Wireframe>   <Join...>

Command: **AMJOIN3D**

> [Mode: **Automatic**
> Output: **3D Polyline**
> **OK**   ]

Select start wire: **[Select F (Figure 2.28).]**
Select wires to join: **[Select G (Figure 2.28).]**
Select wires to join: **[Select H (Figure 2.28).]**
Select wires to join: **[Enter]**
Reverse? Yes/<No>: **NO**

[Standard Toolbar]   [Object Group]

Command: **GROUP**

> [Group name: **W33B**
> Selectable: **Yes**
> **New:**   ]

Select objects for grouping:
Select objects: **[Select D and G (Figure 2.28).]**
Select objects: **[Enter]**

> **[OK]**

To hide the entity groups W33A and W33B, run the AMVISIBLE command. You will use them to produce a swept surface for the top of the body of the camera. See Figure 2.29.

<Surfaces>   <Object Visibility...>

Command: **AMVISIBLE**

> **[Hide
> Select**   ]

Select objects to hide: **GROUP**
Enter group name: **W33A,W33B**
Select objects to hide: **[Enter]**

> **[OK]**

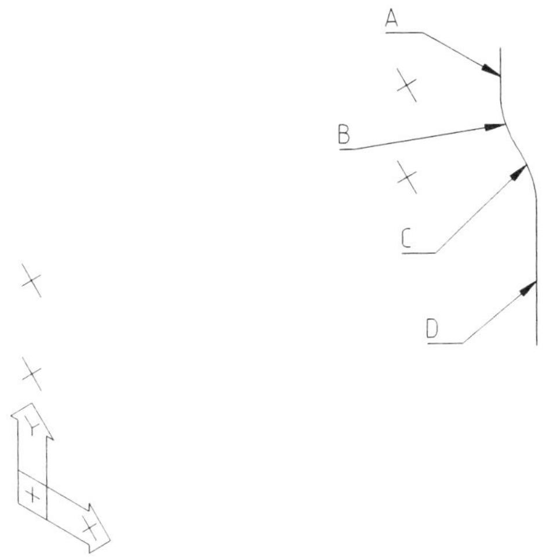

Figure 2.29 Entity groups W33A and W33B hidden

Use the AMJOIN3D command to join the line and arc segments on the screen to become a 3D polyline.

<**Surfaces**>   <**Edit Wireframe**>   <**Join...**>

Command: **AMJOIN3D**

    [Mode: **Automatic**
    Output: **3D Polyline**
    OK   ]

Select start wire: **[Select A (Figure 2.29).]**
Select wires to join: **[Select B, C, and D (Figure 2.29).]**
Select wires to join: **[Enter]**
Reverse? Yes/<No>: **NO**

Change to a new UCS origin with the UCS command.

[**UCS**]   [**Z Axis Vector UCS**]

Command: **UCS**
Origin/ZAxis/3point/OBject/View/X/Y/Z/Prev/Restore/Save/Del/?/<World>: **ZA**
Origin point: ***55,0,50**
Point on positive portion of Z-axis: **@1,0**

In the above command line input, (*55,0,50) means an absolute value of (55,0,50) of the WCS. (@1,0) indicates that the position of Z axis of the new UCS is at (1,0) of the new origin.

Use the LINE command to draw two lines, and the ARC command to draw an arc. See Figure 2.30.

[**Draw**]   [**Line**]

Command: **LINE**
From point: **0,25**
To point: **@95<0**
To point: **[Enter]**

[Draw]   **[Line]**

Command: **LINE**
From point: **0,0**
To point: **@150<0**
To point: **[Enter]**

[Draw]   **[Arc Center Start Angle]**

Command: **ARC**
Center/<Start point>: **C**
Center: **95,-30**
Start point: **@55<0**
Angle/Length of chord/<End point>: **@55<90**

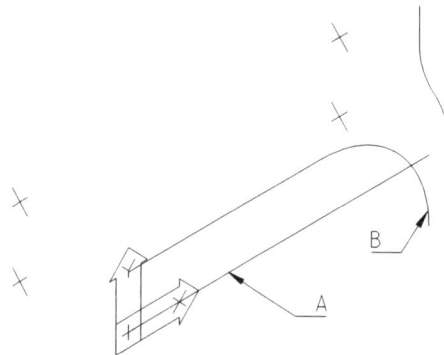

Figure 2.30 Lines and arc drawn

Use the TRIM command to trim the lower portion of the arc. See Figure 2.31.

[Modify]   **[Trim]**

Command: **TRIM**
Select cutting edges: (Projmode = View, Edgemode = No extend)
Select objects: **[Pick the line A (Figure 2.30).]**
Select objects: **[Enter]**
<Select object to trim>/Project/Edge/Undo: **[Select B (Figure 2.30).]**
<Select object to trim>/Project/Edge/Undo: **[Enter]**

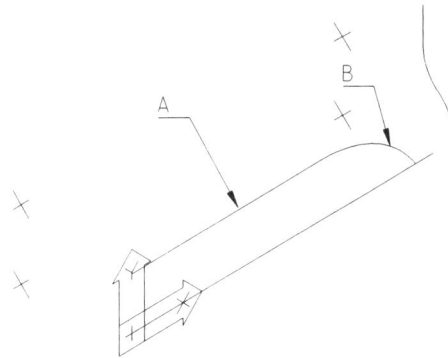

Figure 2.31 Arc trimmed

Join the upper line and the trimmed arc to form a 3D polyline using the AMJOIN3D command.

    \<Surfaces\>   **\<Edit Wireframe\>**   **\<Join...\>**

Command: **AMJOIN3D**

        [Mode: **Automatic**
        Output: **3D Polyline**
        OK   ]

Select start wire: **[Select A (Figure 2.31).]**
Select wires to join: **[Select B (Figure 2.31).]**
Select wires to join: **[Enter]**
Reverse? Yes/\<No\>: **NO**

After joining, refine the 3D polyline by using the AMREFINE3D command to change the number of control points to ten. Then, use the AMFITSPLINE command to convert it to a spline. The number of control points of a polyline affects the accuracy of fitting a polyline to a spline.

    \<Surfaces\>   **\<Edit Wireframe\>**   **\<Refine\>**

Command: **AMREFINE3D**
Select lines or polylines: **[Select A (Figure 2.31).]**
Select lines or polylines: **[Enter]**
Points/\<Tolerance\>: **POINTS**
Points: **10**

    \<Surfaces\>   **\<Edit Wireframe\>**   **\<Spline Fit...\>**

Command: **AMFITSPLINE**
Select wires: **LAST**
Select wires: **[Enter]**

        [Accept the default
        OK   ]

Enter RETURN to continue: **[Enter]**

[OK]

Use the GROUP command to put the entities on the screen into three entity groups, named W34A, W34B, and W34C. They will be used to make a revolved surface and an extruded surface for the right side of the body of the camera. See Figure 2.32.

[Standard Toolbar]   **[Object Group]**

Command: **GROUP**

[Group name: **W34A**
Selectable: **Yes**
New:]

Select objects for grouping:
Select objects: **[Select A (Figure 2.32).]**
Select objects: **[Enter]**

[Group name: **W34B**
Selectable: **Yes**
New:]

Select objects for grouping:
Select objects: **[Select B (Figure 2.32).]**
Select objects: **[Enter]**

[Group name: **W34C**
Selectable: **Yes**
New:]

Select objects for grouping:
Select objects: **[Select C (Figure 2.32).]**
Select objects: **[Enter]**

[OK]

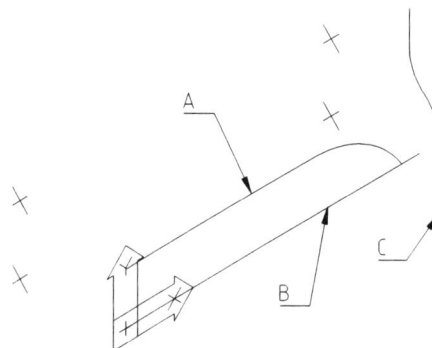

Figure 2.32 Three entity groups -- W34A, W34B, and W34C

You have completed the wireframes for the body of the camera. You created seven entity groups. Group W31 will be used to make a ruled surface for the bottom surface. Group W32 will be used to make a lofted U surface for the left side surface. Groups W33A and W33B will be used to make a swept surface for the top surface, and groups W34A, W34B, and W34C will be used to make a revolved surface and an extruded surface for the right side surface.

Run the AMVISIBLE command to hide all the entities.

Command: **AMVISIBLE**

    [Hide
    All
    OK   ]

2.4 Eyepiece of the Camera

Now we will create the final wireframes for the eyepiece of the video camera. Figure 2.33 shows the completed wireframes of the eyepiece.

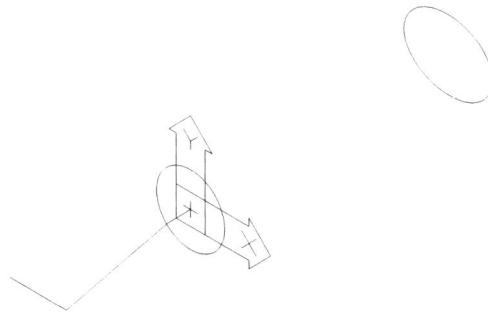

Figure 2.33 Wireframes for the eyepiece of the video camera

Use the UCS command to set the UCS to WORLD. Then, use the 3DPOLY command to create a 3D polyline with three line segments. See Figure 2.34.

    [UCS]   **[World UCS]**

    Command: **UCS**
    Origin/ZAxis/3point/Entity/View/X/Y/Z/Prev/Restore/Save/Del/?/<World>: **W**

    [Draw]   **[3D Polyline]**

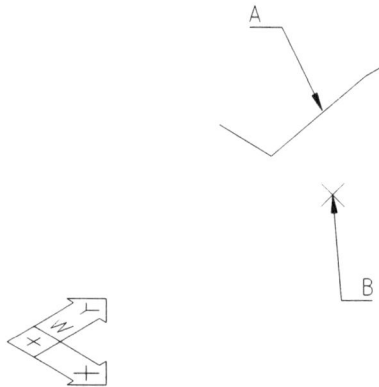

Figure 2.34 3D polyline created

Command: **3DPOLY**
From point: **55,35,98**
Close/Undo/<Endpoint of line>: **@25<0**
Close/Undo/<Endpoint of line>: **@0,45,10**
Close/Undo/<Endpoint of line>: **@10<90**
Close/Undo/<Endpoint of line>: **[Enter]**

Using AutoSurf, you can offset a 3D polyline. Use the AMOFFSET3D command to offset the last 3D polyline for a distance of 20 units. See Figure 2.35.

<Surfaces>   <Create Wireframe>   <Offset Wire>

Command: **AMOFFSET3D**
Offset distance or Through <Through>: **20**
Select 3D polyline to offset: **[Select A (Figure 2.34).]**
Side to offset?: **[Select B (Figure 2.34).]**
Select 3D polyline to offset: **[Enter]**

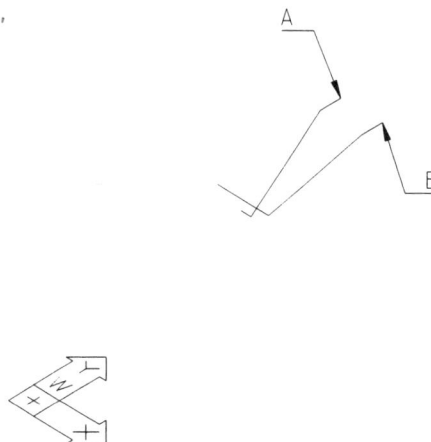

Figure 2.35 3D polyline offset

To verify the offset distance and to see how the AMOFFSET3D command works, run the DIST command to find out the distance between the endpoints of the original and the offset 3D polylines.

[**Inquiry**]   [**Distance**]

Command: **DIST**
First point: **END** of **[Select A (Figure 2.35).]**
Second point: **END** of **[Select B (Figure 2.35).]**
Distance = 20.0000, Angle in XY Plane = 0, Angle from XY Plane = 0
Delta X = 20.0000, Delta Y = 0.0000, Delta Z = 0.0000

The offset polyline is not part of the final model, so erase it using the ERASE command.

[**Modify**]   [**Erase**]

Command: **ERASE**
Select objects: **LAST**
Select objects: **[Enter]**

Set the UCS to rotate about the current X axis for 90°, and move the origin to the far end of the 3D polyline using the UCS command. See Figure 2.36.

[**UCS**]   [**X Axis Rotate UCS**]

Command: **UCS**
Origin/ZAxis/3point/Entity/View/X/Y/Z/Prev/Restore/Save/Del/?/<World>: **X**
Rotation angle about X axis: **90**

[**UCS**]   [**Origin UCS**]

Command: **UCS**
Origin/ZAxis/3point/Entity/View/X/Y/Z/Prev/Restore/Save/Del/?/<World>: **O**
Origin point: **End** of **[Select B (Figure 2.35).]**

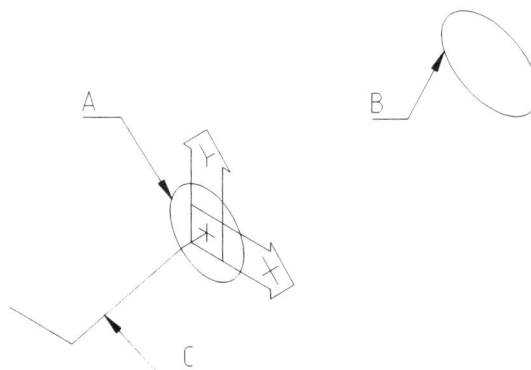

Figure 2.36 Circle and ellipse drawn

Create a circle using the CIRCLE command, and an ellipse using the ELLIPSE command at the new origin. Then, use the MOVE command to translate the ellipse a distance of 80 units to the negative Z direction. See Figure 2.36 again.

[Draw]   **[Circle Center Radius]**

Command: **CIRCLE**
CIRCLE 3P/2P/TTR/<Center point>: **0,0**
Diameter/<Radius>: **15**

[Draw]   **[Ellipse Center]**

Command: **ELLIPSE**
<Axis endpoint 1>/Center: **C**
Center of ellipse: **0,0**
Axis endpoint: **@20<0**
<Other axis distance>/Rotation: **15**

[Modify]   **[Move]**

Command: **MOVE**
Select objects: **LAST**
Select objects: **[Enter]**
<Base point or displacement>/Multiple: **0,0,-115**
Second point of displacement: **[Enter]**

Put the circle and the ellipse in entity group W41, and the 3D polyline in entity group W42 using the GROUP command. The wireframes for making the eyepiece of the camera are completed. To sum up, W41 will be used to create a ruled surface and W42 will be used to create a tubular surface.

[Standard Toolbar]   **[Object Group]**

Command: **GROUP**

    [Group name: **W41**
    Selectable: **Yes**
    New:]

Select objects for grouping:
Select objects: **[Select A and B (Figure 2.36).]**
Select objects: **[Enter]**

    [Group name: **W42**
    Selectable: **Yes**
    New:]

Select objects for grouping:
Select objects: **[Select C (Figure 2.36).]**
Select objects: **[Enter]**

    [OK]

2.5   Completed Wireframes for the Video Camera

You have completed all the wireframes for defining the surface model of the video camera. Now, let us look at what you have done by unhiding all the objects with the AMVISIBLE command. Set the UCS to WORLD using the UCS command. See Figure 2.37.

<Surfaces>   <Object Visibility...>

Command: **AMVISIBLE**

        [Unhide
        All
        OK   ]

[UCS]   [World UCS]

Command: **UCS**
Origin/ZAxis/3point/Entity/View/X/Y/Z/Prev/Restore/Save/Del/?/<World>: **W**

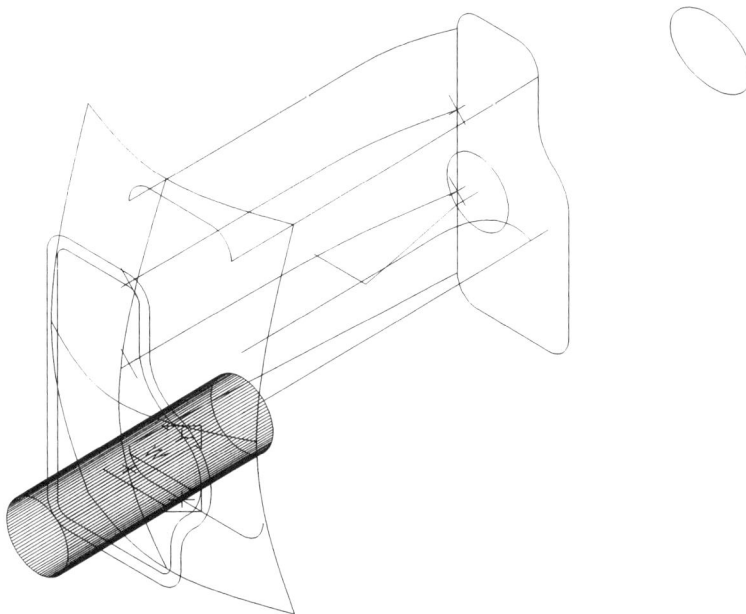

Figure 2.37 All entities drawn in this chapter

The current viewing position is an isometric view. If you want to examine the wireframes from another viewing angle, you can run the AMVIEW command.

<Mtools>   <Mechanical Views>   <Angle>

Command: **AMVIEW**
(Angle = 15)
Angle/Down/eXit/Left/Right/Sketch/Up/<Fit>: **A**

Rotation angle: **10**
(Angle = 10)
Down/eXit/Fit/Left/Right/Sketch/Up/<Angle>: **L**
Angle/Down/eXit/Fit/Right/Sketch/Up/<Left>: **[Enter]**
Angle/Down/eXit/Fit/Right/Sketch/Up/<Left>: **[Enter]**
Angle/Down/eXit/Fit/Right/Sketch/Up/<Left>: **X**

To run the AMVIEW command, you can use the following shortcut keys:

[	Rotates the view to the left.
]	Rotates the view to the right.
=	Rotates the view upward.
-	Rotates the view downward.

Reset the viewport to an isometric view using the accelerator key [8]. Save your drawing before you proceed to the next chapter.

Command: **8**

Command: **SAVE**

2.6   Summary

In this chapter, you have practiced the following AutoSurf commands in 3D wireframe creation:

AMDIRECTION	AMFILLET3D	AMFITSPLINE
AMJOIN3D	AMOFFSET3D	AMREFINE3D
AMUNSPLINE	AMVIEW	AMVISIBLE
AMCHECKFIT	AMSURFVARS	AMVER

For a brief explanation of these commands, refer to the appendix of this book.

In making the wireframes, you created two kinds of wireframes — 3D polylines and splines — and have seen how they differ in terms of continuity. In preparing the wireframes, you learned that wireframes have direction. You also learned how to join wireframes together, how to refine a 3D polyline, how to fit a 3D polyline to become a spline, how to unspline, how to form fillets, and how to offset in 3D space. In addition, you practiced putting entities into groups and controlling their visibility.

In the next chapter, you will complete the 3D surface model of the video camera by constructing NURBS surfaces from the wireframes you created in this chapter.

2.7   Exercise

To enhance your knowledge of preparing wireframes for NURBS surfaces, you will build a set of wireframes for the surface model of a scale model car. The completed set of wireframes is shown in Figure 2.38. You will use these wireframes in the next chapter to complete the model. Figure 3.61 in Chapter 3 shows the completed surface model of the car.

In this section, the steps for making the wireframes are briefly outlined. If you encounter any problems while making the model, you should consult the steps in this chapter or refer to the command reference in the appendix.

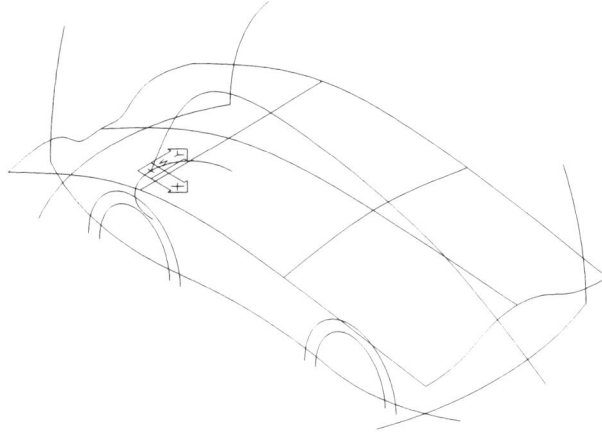

Figure 2.38 Completed wireframe for the model car

To organize the entities into layers, create an additional layer WIRE and set it as the current layer. Create four arcs shown in Figure 2.39 using the following dimensions:

Center	Radius	Start Angle	End Angle
80,210	300	240	285
340,105	200	250	300
180,0	230	150	210
50,0	370	340	20

In additional to the arcs, create a half ellipse with center at (50,0), axis end point at (@50<0), and other axis distance of 48 units

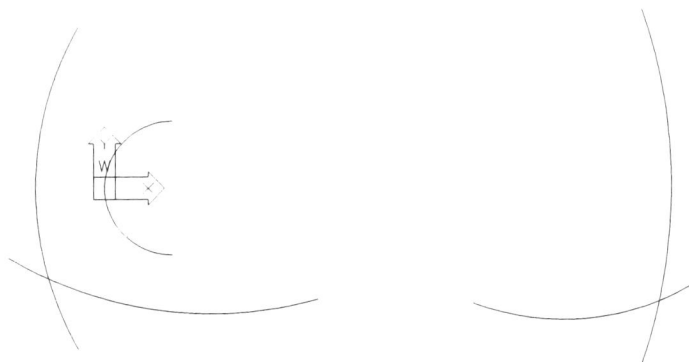

Figure 2.39 Four arcs and one half ellipse created

As shown in Figure 2.40, create a fillet (B) with a radius of 580 units. Then, join together the three arcs (A, B, and C) to form a spline.

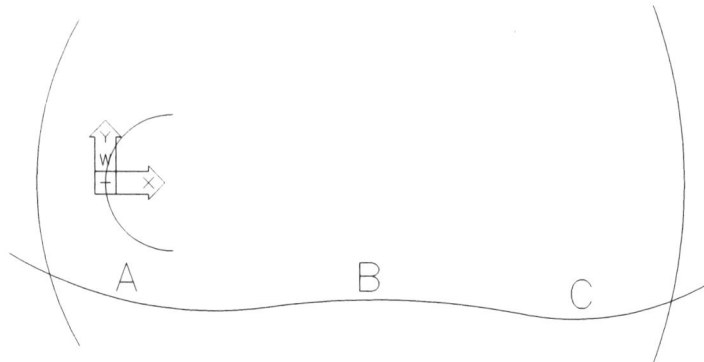

Figure 2.40 Arcs A and C filleted, and arcs A, B, and C joined to become a spline

Put the half ellipse into an entity group called GH, which will be used to create the green house of the model car. Put the remaining entities into a group called BD, which will be used to make the car body.

Use shortcut keys to set to an isometric view and set the orientation of the UCS accordingly. Then, create an arc with a radius of 500 units, a start angle of 165°, and an end angle of 180°. Add this arc to the entity group BD. See Figure 2.41.

Figure 2.41 UCS set and an arc created

Change the UCS orientation. Then, create two more arcs (A and B) as follows, and add them to the entity group BD. See Figure 2.42.

Arc	Radius	Start Angle	End Angle
A	200	140	180
B	300	0	25

Create two half ellipses (C and D) using the following data:

Ellipse	Center	Axis end point	Other axis distance
C	80,0	@55<0	64
D	80,0	@42<0	53

Make a spline E to pass through the points (0,0), (72,113), (210,120), and, (475,20). Add it to the entity group GH.

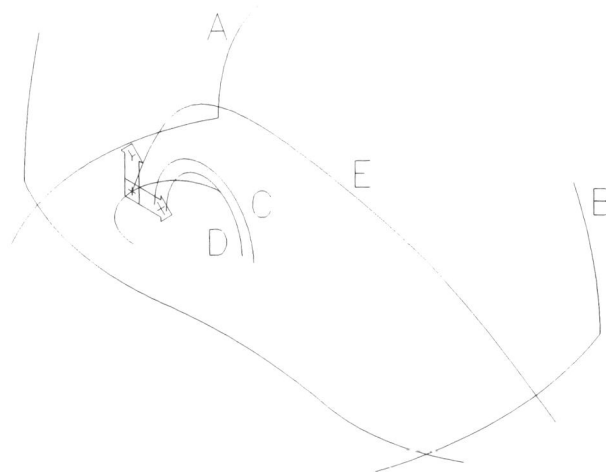

Figure 2.42 Two arcs, two half ellipses, and a spline created

Move the two half ellipses a distance of 100 units in the current Z direction. Then, copy them at a distance of 260<0. See Figure 2.43.

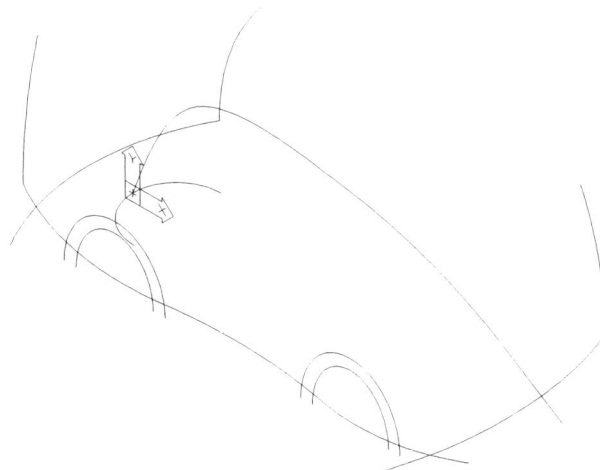

Figure 2.43 Two half ellipses moved and copied

After moving and copying, put the four half ellipses in an entity group called WH, which will be used to create the wheel openings of the car body.

Set the UCS to WORLD. Then, create seven splines. See Figure 2.44.

Spline A passes through the following points:

(-60,-110,24), (97,-110,84), (270,-110,82), and (440,-110,56).

Spline B passes through the following points:

(-60,0,13), (94,0,80), (268,0,93), and (440,0,83).

Spline C passes through the following points:

(-60,110,24), (97,110,84), (270,110,82), and (440,110,56)

Spline D passes through the following points:

(-60,-110,24), (-60,-40,24), (-60,-25,13), (-60,0,13), (-60,25,13), (-60,40,24) and (-60,110,24).

Spline E passes through the following points:

(97,-110,84), (94,0,80), and (97,110,84).

Spline F passes through the following points:

(270,-110,82), (268,0,93), and (270,110,82).

Spline G passes through the following points:

(440,-110,56), (440,-70,61), (440,-30,78), (440,0,83), (440,30,78), (440,70,61), and (440,110,56).

Put these splines into an entity group called TP. They will be used for making the top surface of the model car.

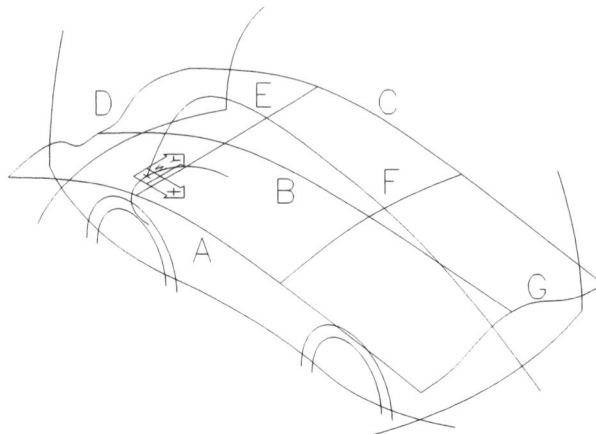

Figure 2.44 Seven splines created

To sum up, you have created a set of wireframes and put them in four entity groups; group BD will be used to make the body; group WH will be used to make the wheel openings; group GH will be used to make the green house; and group TP will be used to make the top surface of the car body. Save your work in a file named CAR.DWG. You will continue this exercise in the next chapter.

Chapter 3
NURBS Surface Modeling

Using the 3D wireframes that you created in the last chapter, you will complete the surface model of the video camera by building a set of 3D NURBS surfaces in this chapter. See Figure 3.1. Similar to the creation of the wireframes, you will build the NURBS surface model in four stages — the frontal part, the rear part, the body portion, and the eyepiece. In the process of making this model, you will learn how to use AutoSurf surface creation and editing commands in engineering application.

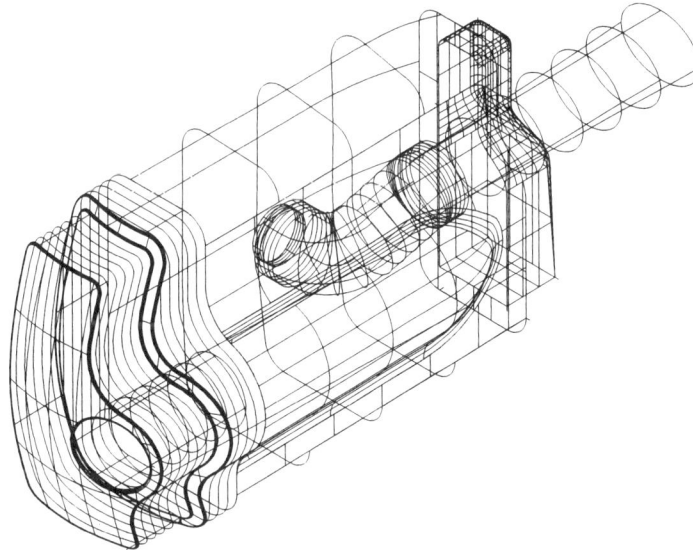

Figure 3.1 Completed NURBS surface model for the video camera

Before you start, check your drawing to ensure that you have set the UCS to WORLD, unhidden all the wireframe entities, and fitted them to the screen in an isometric view.

3.1 Frontal Part of the Camera

Figure 3.2 shows the completed surface model for the frontal part of the video camera. To make this part of the model, you will make a smooth lofted UV surface from two sets of orthogonal splines. Then, you will use a set of 2D wireframe to project and trim this surface. Thus, you will obtain a smooth surface with an irregular boundary. Such a surface is called a trimmed surface. To appreciate how AutoSurf remembers a trimmed surface in memory, you will turn on the display of the base surface. A base surface is the original untrimmed surface. The base surface, together with the trim boundary, form integrated data to represent the required surface. To revert a trimmed surface to its untrimmed state, you can simply remove the trim boundary from the memory.

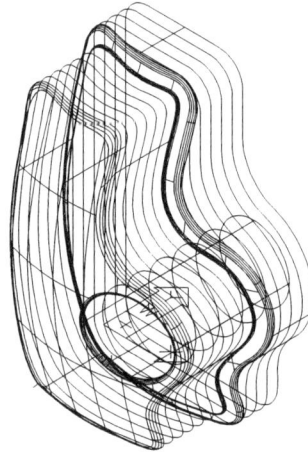

Figure 3.2 Surfaces for the frontal part of the video camera

In general, a large surface requires more memory to store than a small surface. Therefore, you will truncate the base surface of a trimmed surface to its minimum size in order to take up less storage space. However, the removed surface data will be lost permanently. The opposite of truncating is lengthening; you will learn how to lengthen a surface as well.

In addition to using a wireframe to project and trim a surface, you will use a wireframe to project another wireframe on a surface. Then, you will use the resulting wireframe and other wireframe for extrusion.

To proceed on the project, you will offset a surface from an existing surface in order to obtain a pair of surfaces running a constant distance apart.

Smooth surfaces need to be made from smooth wireframes. The frontal part of the camera consists of a number of smooth surfaces that meet. To create these surfaces, you will make smooth surfaces that are larger than required and then trim them. To trim the surfaces, you will perform intersection and filleting. You can regard intersecting as filleting with zero radius or as intersecting, trimming, and rounding up the edges.

After making the required surfaces, you will practice how to break a surface into a number of surfaces that retain the contour and profile of the original surface. You will also join surfaces together.

In older releases of AutoCAD, people used to change the thickness of a 2D wireframe to obtain the elusion of a 3D object. If you have such an object and intend to convert it to an AutoSurf surface, you can do so. In fact, you can convert other objects to surfaces as well.

Create a layer called SURF1 with color yellow, and set it as the current layer. You will put all the NURBS surfaces related to the frontal part of the camera on this layer.

<Data>   **<Layers...>**

Command: **DDLMODES**

Layer	Color
SURF1	**yellow**

Current layer: **SURF1**

With the AMVISIBLE command, hide all the entities except the entity groups W13A and W13B. You will use them to create a lofted UV surface. See Figure 3.3.

<Surfaces>   **<Object Visibility...>**

Command: **AMVISIBLE**

> [**Hide**
> **All**
> **Except**   ]

Select objects to remain visible: **GROUP**
Enter group name: **W13A,W13B**
Select objects to remain visible: **[Enter]**

> [**OK**]

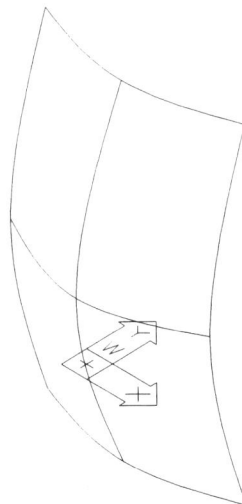

Figure 3.3 Entity groups W13A and W13B unhidden

A lofted UV surface uses two sets of input lines. AutoSurf calls them U-lines and V-lines. These two sets of lines are usually orthogonal to each other. They define the bound and contours of the surface in their directions. Run the AMLOFTUV command. Use the entity group W13A as the U-lines and the entity group W13B as the V-lines.

<Surfaces>   <Create Surface>   <Loft UV>

Command: **AMLOFTUV**
Select U wires: **GROUP**
Enter group name: **W13A**
Select U wires: **[Enter]**
Select V wires: **GROUP**
Enter group name: **W13B**
Select V wires: **[Enter]**

After creating the lofted UV surface, you do not need the six splines anymore. Run the AMVISIBLE command to hide them. Also unhide the entity group W11 for the next operation. See Figure 3.4.

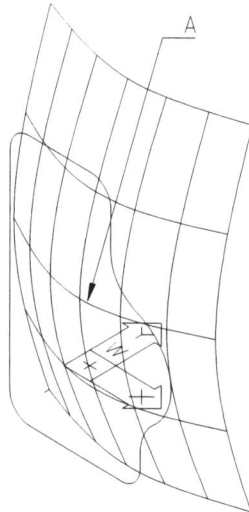

Figure 3.4 Lofted UV surface created and entity group W11 unhidden

<Surfaces>   <Object Visibility...>

Command: **AMVISIBLE**

        [Hide
          Select   **]**

Select objects to hide: **GROUP**
Enter group name: **W13A,W13B**
Select objects to hide: **[Enter]**

        [Unhide
          Select   **]**

Select objects to unhide: **GROUP**

Enter group name: **W11**
Select objects to unhide: **[Enter]**

[**OK**]

When you use an object to create another object, you can choose to retain or delete the original object. In the following operation, you need to keep the source object. Therefore, run the AMSURFVARS command and check the box next to Keep Original. Alternatively, you can set the system variable DELOBJ to zero.

<Surfaces>   <Preferences...>

Command: **AMSURFVARS**

[**Keep Original
OK**   ]

Command: **DELOBJ**
New value for DELOBJ : **0**

Use the AMPROJECT command to project the wireframe W11 onto the newly created lofted UV surface to create a trimmed surface. As a result, this surface possesses its original smooth profile and has an irregular boundary. See Figure 3.5.

<Surfaces>   <Edit Surface>   <Project Trim...>

Command: **AMPROJECT**
Select wires to project: **GROUP**
Enter group name: **W11**
Select wires to project: **[Enter]**
Select target surfaces: **[Select A (Figure 3.4).]**
Select target surfaces: **[Enter]**

[Direction: **Normal to surface**
 Output type: **Trim surface**
 OK   ]

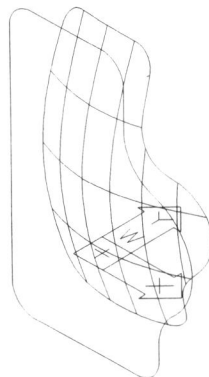

Figure 3.5 Lofted UV surface trimmed by projection of a wireframe

You have produced a smooth lofted UV surface with an irregular boundary. Let us look at the original surface again by using the AMDISPSF command on the trimmed surface.

Change the Normal length to 10 units, the Number of U-lines to 4 and the Number of V-lines to 3. Also click the [Show base surface] button to exhibit the base surface. See Figure 3.6.

<Surfaces>   <Display...>

Command: **AMDISPSF**
Select surfaces: **LAST**
Select surfaces: **[Enter]**

[Persistent Display:
Normal Length: **10**
Number of U-lines: **5**
Number of V-lines: **3**
Temporary Display:
Show base surface: **Yes**
OK   ]

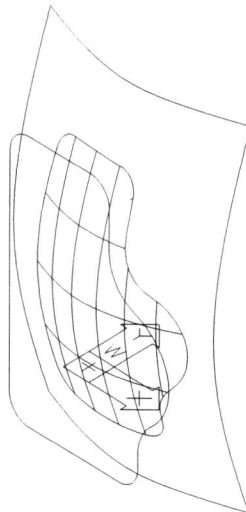

Figure 3.6 Display of the original base surface after trimming

As shown in the dialog box of the AMDISPSF command, there are two types of displays: Persistent display and Temporary display. Normals and UV lines are persistence display. The normal indicates the positive direction of the surface. The UV lines concern the visual appearance of the surface on the screen. Any change to normal length or the number of UV lines will not affect the surface contour.

> **Note**:
> The direction of normal is important if you will use the surface model later for other purposes, such as machining or rendering. If the normal direction is not what you want, you can reverse it by using the AMEDITSF command.

Checking the box next to Show base surface will exhibit the base surface. The display is temporary. The base surface is the original untrimmed surface. AutoSurf database keeps this information to define the smooth surface. You will come to Patches and Control points later in this chapter.

Because the base surface and the trim border of a trimmed surface remain in the memory, you can remove the trim border to revert the trimmed surface to its untrimmed state at any time. If you are not satisfied with the resulting trimmed surface, delete the trim border and trim the surface again.

Use the REDRAW command to refresh the screen. Then, run the AMEDGE command to remove the trim border.

[Standard Toolbar]   **[Redraw View]**

Command: **REDRAW**

<Surfaces>   **<Edit Trim Edges>**   **<Untrim Surface>**

Command: **AMEDGE**
(Output = Polyline)
Copy edge/Output/Show nodes/Untrim/<Extract loop>: **UNTRIM**
Select surfaces: **LAST**
Select surfaces: **[Enter]**

After untrimming, the original untrimmed surface returns. Use the AMPROJECT command to project the wireframe and trim the surface again. This time, take the Vector prompts option and then specify the Y direction. Then, hide the wireframe W11 using the AMVISIBLE command. See Figure 3.7.

<Surfaces>   **<Edit Surface>**   **<Project Trim...>**

Command: **AMPROJECT**
Select wires to project: **GROUP**
Enter group name: **W11**
Select wires to project: **[Enter]**
Select target surfaces: **[Select A (Figure 3.4).]**
Select target surfaces: **[Enter]**

[Direction: **Vector prompts**
Output type: **Trim surface**
OK   ]

Viewdir/Wire/X/Y/Z/<Start point>: **Y**

<Surfaces>   **<Object Visibility...>**

Command: **AMVISIBLE**

> [**Hide**
> **Select**   ]

Select objects to hide: **GROUP**
Enter group name: **W11**
Select objects to hide: **[Enter]**

> [**OK**]

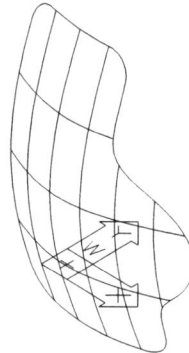

Figure 3.7 Surface trimmed again and group W11 hidden

Refer back to Figure 3.6. You will find that the base surface needed to define the contour and profile of the trimmed surface can be smaller. Therefore, you can truncate it to a smaller size. When the base surface becomes smaller, the file size will be smaller too. However, you will lose the surface data of the truncated away portion permanently.

To reduce the base surface, use the AMEDITSF command.

> **<Surfaces>**   **<Edit Surface>**   **<Truncate>**

Command: **AMEDITSF**
Select surfaces: **LAST**
Select surfaces: **[Enter]**

> [Truncate: **Yes**
> **OK**   ]

After truncating, you can view the base surface again with the AMDISPSF command to verify the change. You will notice that the base surface is much smaller. It is now just a little larger than the trimmed surface. See Figure 3.8.

> **<Surfaces>**   **<Display...>**

Command: **AMDISPSF**
Select surfaces: **LAST**
Select surfaces: **[Enter]**

[Temporary Display:
Show base surface: **Yes**
OK   ]

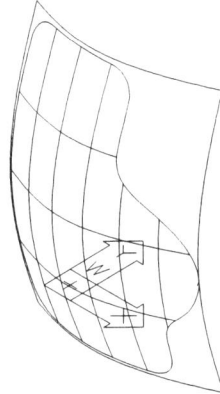

Figure 3.8 Base surface truncated

Use the REDRAW command to clear the temporary display of the base surface. Then, use the GROUP command to put this surface in an entity group named S11.

[Standard Toolbar]   **[Redraw View]**

Command: **REDRAW**

[Standard Toolbar]   **[Object Group]**

Command: **GROUP**

[Group name: **S11**
Selectable: **Yes**
New:]

Select objects for grouping:
Select objects: **LAST**
Select objects: **[Enter]**

[OK]

Unhide the wireframe W12 using the AMVISIBLE command. You will use this wireframe to project a wireframe on the lofted UV surface. See Figure 3.9.

<Surfaces>   **<Object Visibility...>**

Command: **AMVISIBLE**

[**Unhide**
  Select   ]

Select objects to unhide: **GROUP**
Enter group name: **W12**
Select objects to unhide: **[Enter]**

   [OK]

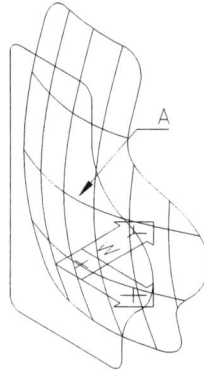

Figure 3.9 Wireframe W12 unhidden

In addition to using a wireframe to project and trim a surface, you can project a wireframe on a surface to obtain another wireframe. Run the AMPROJECT command to output a polyline. See Figure 3.10.

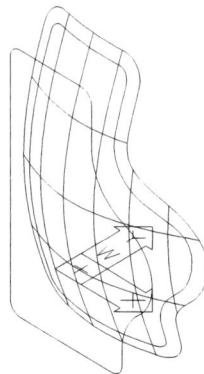

Figure 3.10 Wireframe created by projection

<Surfaces>   <Create Wireframe>   <Project Wire...>

Command: **AMPROJECT**
Select wires to project: **GROUP**
Enter group name: **W12**
Select wires to project: **[Enter]**
Select target surfaces: **[Select A (Figure 3.9).]**
Select target surfaces: **[Enter]**

[Direction: **Vector prompts**
Output type: **Polyline**
OK   ]

Viewdir/Wire/X/Y/Z/<Start point>: **Y**

After projection, change the projected wireframe to layer WIRE1.

<Edit>   **<Properties...>**

Select objects: **LAST**
Select objects: **[Enter]**

[Properties
Layer...   **WIRE1**
OK   ]

Hide the original wireframe W12 using the AMVISIBLE command.

<Surfaces>   **<Object Visibility...>**

Command: **AMVISIBLE**

[**Hide**
Select   ]

Select objects to hide: **GROUP**
Enter group name: **W12**
Select objects to hide: **[Enter]**

[**OK**]

To demonstrate that a projected wireframe can be used for surface creation, run the AMEXTRUDESF command on this wireframe to obtain an extruded surface. See Figure 3.11.

<Surfaces>   **<Create Surface>**   **<Extrude>**

Command: **AMEXTRUDESF**
Select wires: **LAST**
Select wires: **[Enter]**
Direction: Viewdir/Wire/X/Y/Z/<Start point>: **Y**
Distance: **-35**
Flip/<Accept>: **ACCEPT**
Taper angle: **0**

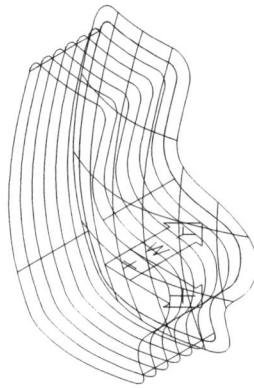

Figure 3.11 Surface of extrusion created

You have, so far, created two NURBS surfaces: a trimmed lofted UV surface and an extruded surface. To continue, use the AMOFFSETSF command to offset the lofted UV surface to derive another surface. An offset surface is similar to the original surface in contour and shape but has different dimensions. See Figure 3.12.

<Surfaces>   <Create Surface>   <Offset>

Command: **AMOFFSETSF**
Select surfaces: **GROUP**
Enter group name: **S11**
Select surfaces: **[Enter]**
Offset distance: **25**

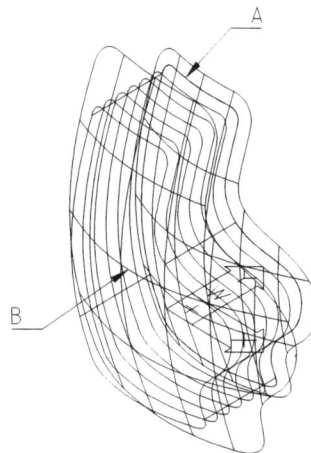

Figure 3.12 Offset surface created

In making a surface model for a product, it is a usual practice to produce a number of smooth surfaces that are large enough to intersect each other, and then trim the unwanted portion.

As you can see, the extruded surface and the offset surface are large enough to intersect. To trim the unwanted part, you will use the AMINTERSF command. See Figure 3.13. When selecting the objects for intersection, you should select where you want to retain.

\<Surfaces>   **\<Create Wireframe>**   **\<Intersect Wire...>**

Command: **AMINTERSF**
Select first surface: **[Select A (Figure 3.12).]**
Select second surface: **[Select B (Figure 3.12).]**

       [Trim surface:
       First surface: **Yes**
       Second surface: **Yes**
       OK   ]

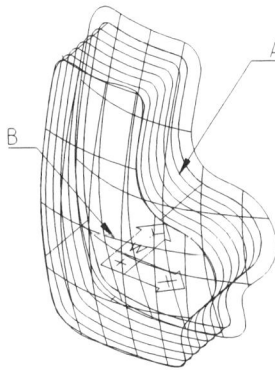

Figure 3.13 Intersecting surfaces trimmed

After the intersecting process, the two trimmed surfaces form a sharp edge. Use the AMFILLETSF command to round off this edge with a constant fillet radius of 1 unit. See Figure 3.14.

\<Surfaces>   **\<Create Surface>**   **\<Fillet...>**

Command: **AMFILLETSF**
Select first surface: **[Select A (Figure 3.13).]**
Select second surface: **[Select B (Figure 3.13).]**

       [Trim:
       First surface: **Yes**
       Second surface: **Yes**
       Fillet Radius: **1**
       OK   ]

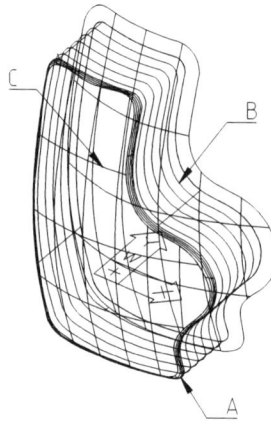

Figure 3.14 Edge filleted

Run the GROUP command to put the fillet surface, the extruded surface, and the offset surface in three entity groups S12, S13, and S14b, respectively.

[Standard Toolbar]   **[Object Group]**

Command: **GROUP**

> [Group name: **S12**
> Selectable: **Yes**
> **New:**   ]

Select objects for grouping:
Select objects: **[Select A, the fillet surface (Figure 3.14).]**
Select objects: **[Enter]**

> [Group name: **S13**
> Selectable: **Yes**
> **New:**   ]

Select objects for grouping:
Select objects: **[Select B, the extruded surface (Figure 3.14).]**
Select objects: **[Enter]**

> [Group name: **S14**
> Selectable: **Yes**
> **New:**   ]

Select objects for grouping:
Select objects: **[Select C, the offset surface (Figure 3.14).]**
Select objects: **[Enter]**

> **[OK]**

Run the COPY command to copy the lofted UV surface a distance of 100<0. Then, use the AMVISIBLE command to hide the entity groups S11, S12, S13, S14, and the projected wireframe that is used for extrusion.

[Modify]   **[Copy Object]**

Command: **COPY**
Select objects: **GROUP**
Enter group name: **S11**
Select objects: **[Enter]**
<Base point or displacement>/Multiple: **100<0**
Second point of displacement: **[Enter]**

<Surfaces>   **<Object Visibility...>**

Command: **AMVISIBLE**

  **[Hide
    Select   ]**

Select objects to hide: **GROUP**
Enter group name: **S11,S12,S13,S14**
Select objects to hide: **[Select the projected wireframe used for extrusion.]**
Select objects to hide: **[Enter]**

  [OK]

On the screen, you should have only one surface left -- the copied surface. Because this surface is not part of the final model, you will try some editing commands on it.

Opposite to truncating a surface, you will lengthen this surface. Run the AMLENGTHEN command.

<Surfaces>   **<Edit Surface>**   **<Lengthen>**

Command: **AMLENGTHEN**
(Base edge=Single, Extension=Percent, Method=Parabolic, Value= 110%)
Base edge/Extension/Method/Value/<Select surface edge>: **[Select the upper edge of
                                the surface shown on the screen.]**
Selected edge must be an untrimmed base edge.
Base edge/Extension/Method/Value/<Select surface edge>: **[Enter]**

As you can see, the request to lengthen the trimmed surface is rejected. Why? To answer this question, it is necessary to reiterate that a trimmed surface consists of a base surface and a trim border. You cannot lengthen a surface at the trim border. If you really want to lengthen a trimmed surface, you should first untrim the surface. Then, you can lengthen at the border of the base surface.

To proceed, run the AMEDGE command to remove the trimmed border, and then use the AMLENGTHEN command again.

<Surfaces>   **<Edit Trim Edges>**   **<Untrim Surface>**

Command: **AMEDGE**
(Output = Polyline)
Copy edge/Output/Show nodes/Untrim/<Extract loop>: **UNTRIM**
Select surfaces: **[Select the copied surface.]**
Select surfaces: **[Enter]**

While lengthening the surface, you can choose to keep the original surface or not. If you do not want the original surface after lengthening, set the system variable DELOBJ to 1.

Command: **DELOBJ**
New value for DELOBJ: **1**

<Surfaces>   <Edit Surface>   <Lengthen>

Command: **AMLENGTHEN**
(Base edge=Single, Extension=Percent, Method=Parabolic, Value = 110%)
Base edge/Extension/Method/Value/<Select surface edge>: **B**
Single/<All>: **S**
Base edge/Extension/Method/Value/<Select surface edge>: **E**
Percent/<Delta>: **D**
Delta <1>: **40**
(Base edge = Single, Extension = Delta, Method = Parabolic, Value = 40)
Base edge/Extension/Method/Value/<Select surface edge>: **[Select the upper edge of the base surface.]**
Base edge/Extension/Method/Value/<Select surface edge>: **[Enter]**

The surface is now lengthened. To continue practicing, you will edit its contour and profile, break it into two pieces, and then rejoin the pieces.

Editing an existing surface will take a few steps. First, you should check the number of patches and control points, and adjust the number of U and V patches, if necessary. Then, you have to set the span of the surface. The span defines a circular area that will be deformed when you pull a control point. After doing all the preparative work, you will use the AutoCAD grip facility to pick and pull a control point to edit the surface.

You are going to refine this lengthened surface. Before you do so, run the AMDISPSF command to look at the Patch normals and Control points.

<Surfaces>   <Display...>

Command: **AMDISPSF**
Select surfaces: **LAST**
Select surfaces: **[Enter]**

             [Temporary Display:
               Show Patch Normals: **Yes**
               Show Control Points: **Yes**
               OK   ]

The Patch normals indicate the patches that define the surface contour. The Control points of a spline surface behave similarly to the control points of a spline. They control the contour of the surface. You need more patches and control points for complex irregular contours, but not for a simple smooth surface like this one. Change the number of patches by refining the surface.

Refresh the screen with the REDRAW command.

[Standard Toolbar]   [Redraw View]

Command: **REDRAW**

To refine the UV patches of a surface, apply the AMREFINESF command.

<Surfaces>   **<Edit Surface>**   **<Refine>**

Command: **AMREFINESF**
Select surfaces: **LAST**
Select surfaces: **[Enter]**
Uv patches/<Tolerance>: **UV**
U patches: **6**
V Patches: **4**

To see the patches and normals after refining, use the AMDISPSF command again. You will find that there are fewer patches and control points. See Figure 3.15.

<Surfaces>   **<Display...>**

Command: **AMDISPSF**
Select surfaces: **LAST**
Select surfaces: **[Enter]**

[Temporary Display:
  Show Patch Normals: **Yes**
  Show Control Points: **Yes**
  OK   ]

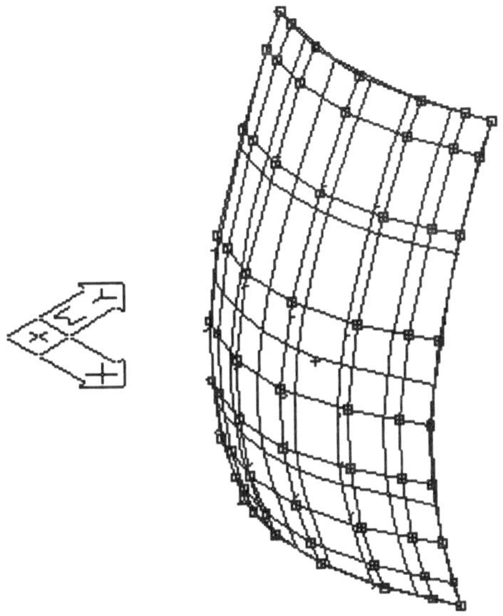

Figure 3.15 Patches and control points shown temporarily

The control points and patches govern the contour and shape of the surface. To edit the contour of a surface, you will move the control points by manipulating the grips.

Each grip point on the surface affects a circular zone of area. AutoSurf calls this area SPAN. Before you can edit a surface, you should set the span radius with the AMEDITSF command.

```
Command: AMEDITSF
Select surfaces: LAST
Select surfaces: [Enter]

       [Grips
         Span:       30
         Preview                        ]

Enter RETURN to continue: [Enter]

       [OK]
```

In the preview, you can see a small, red circle. This circle defines an area that will deform when a grip point is moved.

After exiting the command, set the grips to 1 to enable grips manipulation.

```
Command: GRIPS
New value for GRIPS: 1
```

Select the object to elicit the grip points. Pick the centermost grip point and move it a distance of 40 units in the 270° direction. See Figure 3.16.

```
Command:
** STRETCH **
<Stretch to point>/Base point/Copy/Undo/eXit: @40<270
```

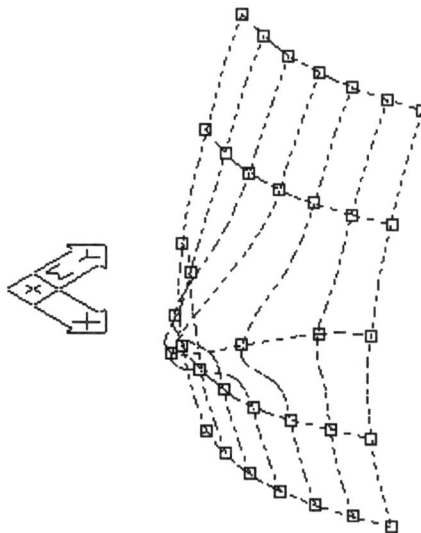

Figure 3.16 Surface edited by pulling the central control point

After you have pulled a grip point, a circular portion of the surface deforms. The radius of this deformed circular area is the span that you have set. To see the effect of editing, perform shading or rendering by using the SHADE or RENDER command.

Command: **SHADE**

After shading or rendering, perform the REGEN command to restore the display.

Command: **REGEN**

You can break a NURBS surface into separate pieces, and yet retain the original contours. Run the AMBREAK command to break the surface into two pieces.

<Surfaces>   <Edit Surface>   <Break>

Command: **AMBREAK**
Select surface: **[Select the upper edge of the surface (Figure 3.16).]**
(Percent = 50%)
Flip/Reposition/<Break u>: **[Enter]**

In order to see the effect of breaking, move the surfaces apart with the MOVE command. See Figure 3.17.

[Modify]   **[Move]**

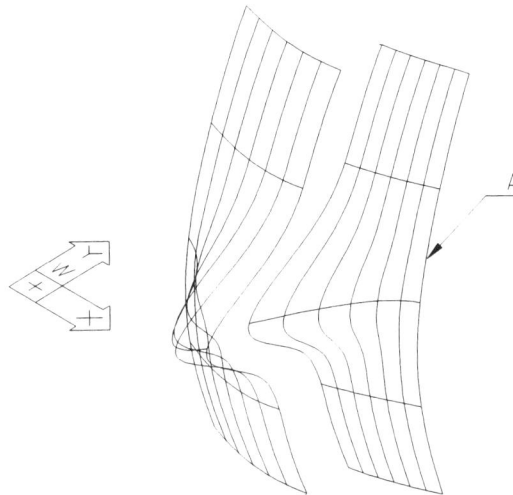

Figure 3.17 Broken surfaces moved apart

Command: **MOVE**
Select objects: **[Select the right edge of the surface (Figure 3.16).]**
Select objects: **[Enter]**
<Base point or displacement>/Multiple: **20<90**
Second point of displacement: **[Enter]**

Opposite to breaking, you can join two surfaces together. Move the broken surfaces back to their original position using the MOVE command, and join them together using the AMJOINSF command.

[Modify]   **[Move]**

Command: **MOVE**
Select objects: **[Select A (Figure 3.17).]**
Select objects: **[Enter]**
<Base point or displacement>/Multiple: **20<270**
Second point of displacement: **[Enter]**

<Surfaces>   **<Edit Surface>**   **<Join>**

Command: **AMJOINSF**
Select surfaces to join: **[Select the left surface (Figure 3.17).]**
Select surfaces to join: **[Select the right surface (Figure 3.17).]**
Select surfaces to join: **[Enter]**

As mentioned earlier, this surface is not part of the final model. You used it only to practice various editing commands. Erase it using the ERASE command.

[Modify]   **[Erase]**

Command: **ERASE**
Select objects: **LAST**
Select objects: **[Enter]**

After trying a number of surface editing commands, you will come back to the frontal part of the camera and produce more surfaces.

Set the system variable DELOBJ to 0. Then, use the AMVISIBLE command to unhide the entity groups S11 and W11. S11 is the lofted UV surface, and W11 is a wireframe that you have used to project and trim that surface. See Figure 3.18.

Command: **DELOBJ**
New value for DELOBJ: **0**

<Surfaces>   **<Object Visibility...>**

Command: **AMVISIBLE**

    [Unhide
    Select   ]

Select objects to unhide: **GROUP**
Enter group name: **S11,W11**
Select objects to unhide: **[Enter]**

    [OK]

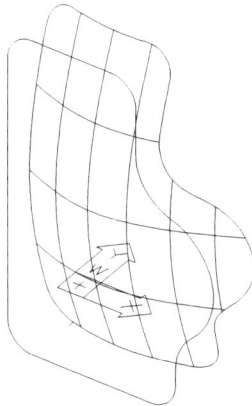

Figure 3.18 Surface S11 and wireframe W11 unhidden

Run the AMEXTRUDESF command to extrude the wireframe W11 for a distance of 40 units in the positive Y direction. See Figure 3.19.

<Surfaces>   <Create Surface>   <Extrude>

Command: **AMEXTRUDESF**
Select wires: **GROUP**
Enter group name: **W11**
Select wires: **[Enter]**
Direction: Viewdir/Wire/X/Y/Z/<Start point>: **Y**
Distance: **40**
Flip/<Accept>: **ACCEPT**
Taper angle: **0**

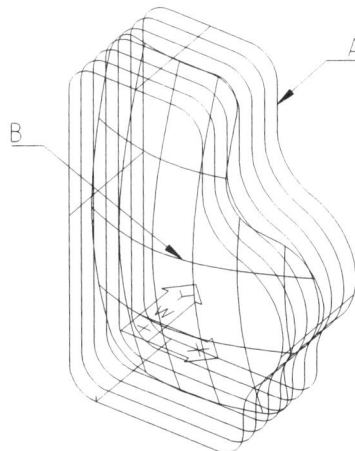

Figure 3.19 Wireframe W11 extruded to form a surface

Use the AMFILLETSF command to create a trimmed fillet edge of constant radius 1 unit between the lofted UV surface S11 and the newly extruded surface. When using the

AMFILLETSF command, be careful to select the central part of the surface S11 and the far end of the newly extruded surface. When you fillet intersecting surfaces, you can, optionally, perform trimming automatically. See Figure 3.20.

\<Surfaces\>   \<Create Surface\>   \<Fillet...\>

Command: **AMFILLETSF**
Select first surface: **[Select A (Figure 3.19).]**
Select second surface: **[Select B (Figure 3.19).]**

```
[Trim:
  First surface:      Yes
  Second surface:     Yes
  Fillet Radius:      1
  OK                        ]
```

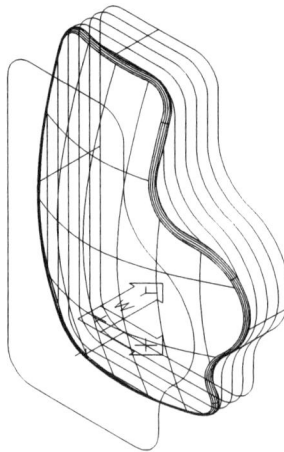

Figure 3.20 Filleting and trimming done in one operation

Wireframe W11 is not needed any more. Execute the AMVISIBLE command to hide it from view. Also, unhide the entity group S13, which is the extruded surface created from the projected wireframe.

Run the AMFILLETSF command to create a constant fillet radius of 2 units between the unhidden surface S13 and the surface S11. While filleting, select what you want to keep. See Figure 3.21.

\<Surfaces\>   \<Object Visibility...\>

Command: **AMVISIBLE**

```
[Hide
  Select                    ]
```

Select objects to hide: **GROUP**
Enter group name: **W11**
Select objects to hide: **[Enter]**

[Unhide
  Select   ]

Select objects to unhide: **GROUP**
Enter group name: **S13**
Select objects to unhide: **[Enter]**

  [OK]

Command: **AMFILLETSF**
Select first surface: **[Select A (Figure 3.21).]**
Select second surface: **[Select B (Figure 3.21).]**

    [Trim:
      First surface: **Yes**
      Second surface: **Yes**
      Fillet Radius: **1**
      OK   ]

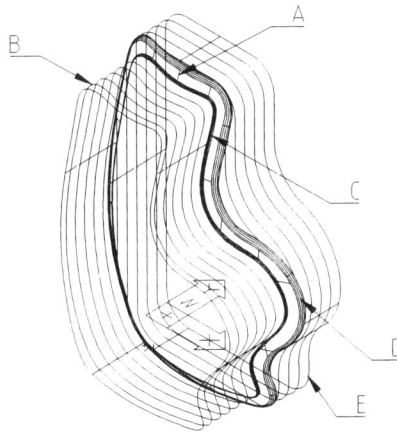

Figure 3.21 Edges filleted and trimmed

After filleting, hide all the surfaces on the screen. Then, unhide the entity groups W14 and S14. W14 is a 2D circle with thickness. S14 is the offset surface. See Figure 3.22.

    <Surfaces>   <Object Visibility...>

Command: **AMVISIBLE**

    [Hide
      Select   ]

Select objects to hide: **GROUP**
Enter group name: **S11,S13**
Select objects to hide: **[Select C, D, and E (Figure 3.21).]**
Select objects to hide: **[Enter]**

    [Unhide

Select   ]

Select objects to unhide: **GROUP**
Enter group name: **W14,S14**
Select objects to unhide: **[Enter]**

   [OK]

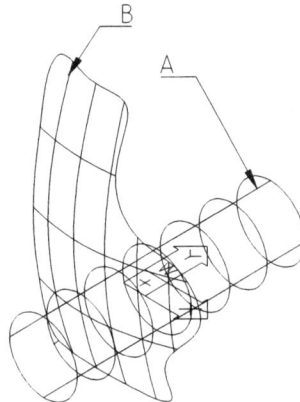

Figure 3.22 AutoCAD entity converted to AutoSurf entity

You can convert some AutoCAD entities to AutoSurf surfaces. To reiterate, entity group W14 is a wireframe with thickness, not a surface. Run the ACAD2SF command on it. Then, change its layer to SURFACE. See Figure 3.22 again.

   <Surfaces>   <Create Surface>   <From ACAD>

   Command: **AM2SF**
   Face/<Objects>: **O**
   Select objects: **GROUP**
   Enter group name: **W14**
   Select objects: **[Enter]**
   Converting: 1 of 1

   Edit>   <Properties...>

   Select objects: **LAST**
   Select objects: **[Enter]**

   [Properties
   Layer...   SURF1
   OK   ]

Because DELOBJ is 0, the circle with thickness remains after a cylindrical surface is created by the AM2SF command. This circle is no longer needed. Run the ERASE command to erase it.

   [Modify]   [Erase]

Command: **ERASE**
Select objects: **GROUP**
Enter group name: **W14**
Select objects: **[Enter]**

To complete the frontal part of the camera, trim and fillet the cylindrical surface with the offset surface by using the AMFILLETSF command. See Figure 3.23.

<Surfaces>   **<Create Surface>**   **<Fillet...>**

Command: **AMFILLETSF**
Select first surface: **[Select A (Figure 3.22).]**
Select second surface: **[Select B (Figure 3.22).]**

        [Trim:
         First surface: **Yes**
         Second surface: **Yes**
         Fillet Radius: **1**
         OK   ]

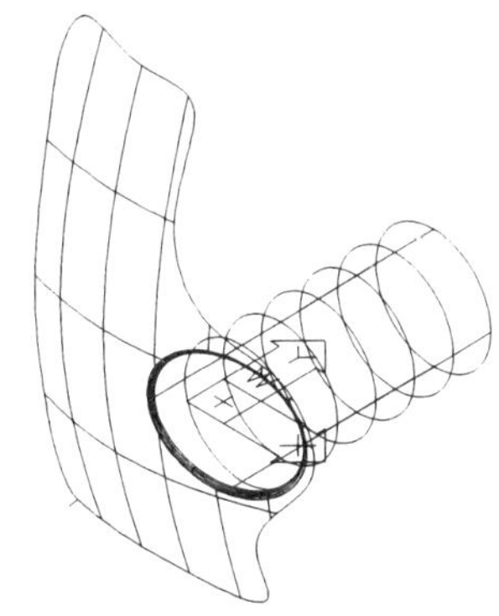

Figure 3.23 Converted entity filleted and trimmed

Use the AMVISIBLE command to unhide all the surfaces. The frontal part of the video camera is completed. Check your work against Figure 3.2.

<Surfaces>   **<Object Visibility...>**

Command: **AMVISIBLE**

        [Unhide
         Select   ]

Select objects to unhide: **GROUP**
Enter group name: **S***
Select objects to unhide: **[Select the two fillet surfaces and one extruded surface on the screen.]**
Select objects to unhide: **[Enter]**

        [OK]

3.2   Rear Part of the Camera

Figure 3.24 shows the surface model for the rear part of the video camera. In making this part of the model, you will create a planar surface, practice converting AutoCAD objects to surfaces, create variable fillets, create a corner surface from three intersecting fillets, and blend surfaces together. In addition, you will copy the edges of surfaces to form wireframes, and use these wireframes to trim another surface.

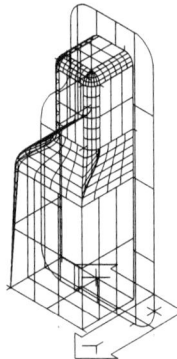

Figure 3.24   Surfaces for the rear part of the video camera

Make a new layer called SURF2. Set its color to red, and set it as the current layer. Also turn off the layer SURF1.

<Data>   **<Layers...>**

Command: **DDLMODES**

Layer	Color	
SURF1		**Off**
SURF2	**red**	

Current layer: **SURF2**

The wireframe used for this part of the model is the entity group W21. Use the AMVISIBLE command to unhide this group. See Figure 3.25.

<Surfaces>   **<Object Visibility...>**

Command: **AMVISIBLE**

    [Unhide
     Select   ]

Select objects to unhide: **GROUP**
Enter group name: **W21**
Select objects to unhide: **[Enter]**

    [OK]

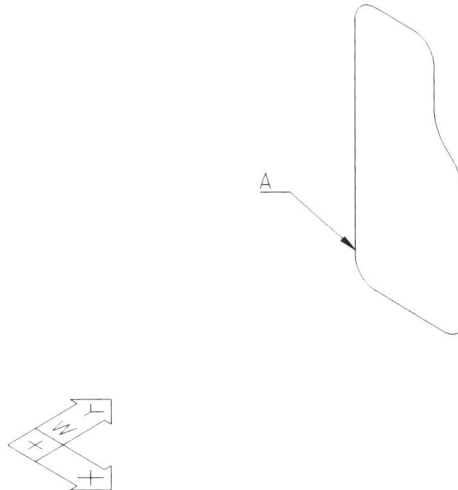

Figure 3.25 Wireframe for the rear part of the video camera

Set the origin of the UCS to a new position. See Figure 3.26.

[UCS]   **[Origin UCS]**

Command: **UCS**
Origin/ZAxis/3point/Entity/View/X/Y/Z/Prev/Restore/Save/Del/?/<World>: **O**
Origin point: **END** of **[Select A (Figure 3.25).]**

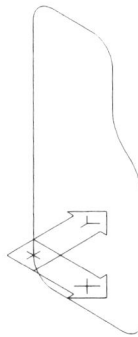

Figure 3.26 UCS origin set to a new position

To see once more how AutoCAD objects can be used to form AutoSurf surfaces, create two solid boxes and then convert them to surfaces.

To delete the original object after conversion, set the system variable DELOBJ to 1.

Command: **DELOBJ**
New value for DELOBJ : **1**

Use the BOX command to create two solid boxes. Then, convert them into two sets of surfaces using the AM2SF command. See Figure 3.27.

Command: **BOX**
Center/<Corner of box>: **10,0**
Cube/Length/<other corner>: **@35,25,50**

Command: **BOX**
Center/<Corner of box>: **10,0,70**
Cube/Length/<other corner>: **@20,20,30**

Command: **AM2SF**
Face/<Objects>: **O**
Select objects: **[Select the two boxes.]**
Select objects: **[Enter]**
Converting: 1 of 2 .....

Surfaces are represented by UV display lines. Sometimes, too many UV display lines makes object selection difficult. To reduce the number of UV display lines, run the AMDISPSF command. See Figure 3.27.

<Surfaces>   <Display...>

Command: **AMDISPSF**
Select surfaces: **[Select all the surfaces on the screen.]**
Select surfaces: **[Enter]**

[Persistent Display:
  Number of U-lines: **2**
  Number of V-lines: **1**
  OK   ]

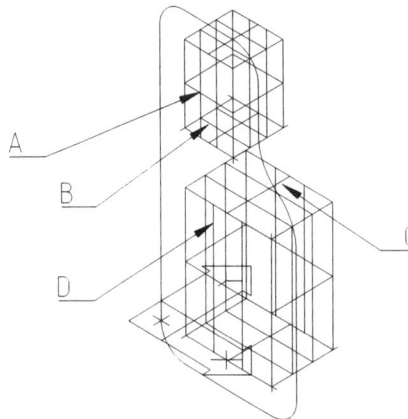

Figure 3.27 Two solid boxes created and converted to surfaces

Four of the surfaces indicated in Figure 3.27 are not required. Use the ERASE command to delete them. Then, set the viewing position using the VPOINT command. See Figure 3.28.

[Modify]   **[Erase]**

Command: **ERASE**
Select objects: **[Select A, B, C, and D (Figure 3.27).]**
Select objects: **[Enter]**

<View>   **<3D Viewpoint>**   **<Vector>**

Command: **VPOINT**
*** Switching to the WCS ***
Rotate/<View point>:**R**
Enter angle in XY plane from X axis: **45**
Enter angle from XY plane: **35**
*** Returning to the UCS ***
Regenerating drawing.

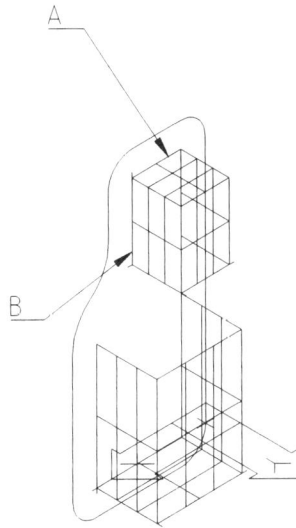

Figure 3.28 New viewing position

You have to round the edges and corners of the upper set of surfaces. To round the edges, use the AMFILLETSF command. To round the corners, use the AMCORNER command.

Use the AMFILLETSF command on two surfaces. See Figure 3.29.

<Surfaces>   **<Create Surface>**   **<Fillet...>**

Command: **AMFILLETSF**
Select first surface: **[Select A (Figure 3.28.]**
Select second surface: **[Select B (Figure 3.28).]**

[Trim:
First surface: **Yes**
Second surface: **Yes**
Fillet Radius: **4**
OK   ]

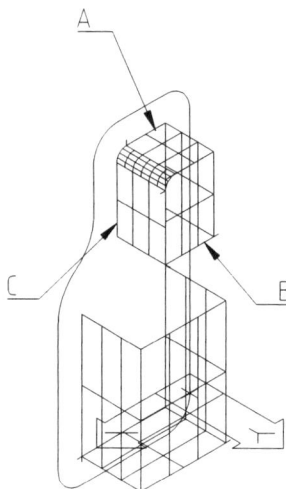

Figure 3.29 One edge filleted

The two surfaces that were filleted are trimmed. They become trimmed surfaces. As explained earlier, a trimmed surface consists of the trim border and the original untrimmed base surface in the database. To proceed filleting these trimmed surfaces with other surfaces to form a corner, you have to use the fillet type Base surface to obtain a fillet running at full length of the original base surface.

Run the AMFILLETSF command twice using the Base surface option. See Figure 3.30.

<Surfaces>   <Create Surface>   <Fillet...>

Command: **AMFILLETSF**
Select first surface: **[Select A (Figure 3.29.]**
Select second surface: **[Select B (Figure 3.29).]**

      [Trim:
      First surface: **Yes**
      Second surface: **Yes**
      Fillet Type:
      Base surface: **Yes**
      Fillet Radius: **4**
      OK   ]

<Surfaces>   <Create Surface>   <Fillet...>

Command: **AMFILLETSF**
Select first surface: **[Select B (Figure 3.29.]**
Select second surface: **[Select C (Figure 3.29).]**

      [Trim:
      First surface: **Yes**
      Second surface: **Yes**
      Fillet Type:
      Base surface: **Yes**

Fillet Radius: **4**

OK   ]

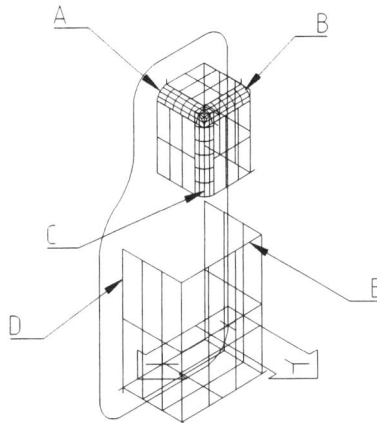

Figure 3.30 Two edges filleted to the base surfaces

Examine your drawing carefully. The three fillet surfaces should run the full lengths of the original base surfaces and should intersect each other. If you do not, you probably have not used the Base surface option. In that case, you need to undo the fillets and create them again.

Once you have ensured that the fillets are correct, round off the corner by using the AMCORNER command. See Figure 3.31.

<Surfaces>   **<Create Surface>**   **<Corner Fillet>**

Command: **AMCORNER**
(Trim = Yes)
Trim/<Select first fillet surface>: **[Select A (Figure 3.30).]**
Trim/<Select second fillet surface>: **[Select B (Figure 3.30).]**
Trim/<Select third fillet surface>: **[Select C (Figure 3.30).]**

In making fillets, you can choose to have a fillet of changing radius — variable fillet. Run the AMFILLETSF command to create a linear variable fillet on the lower set of surfaces. The start radius at the top is 4 units. The end radius at the bottom is 0 units. The radius of a linear variable fillet changes linearly from one set value to another set value. See Figure 3.31.

<Surfaces>   **<Create Surface>**   **<Fillet...>**

Command: **AMFILLETSF**
Select first surface: **[Select D (Figure 3.30.]**
Select second surface: **[Select E (Figure 3.30).]**

        [Trim:
          First surface: **Yes**
          Second surface: **Yes**

Fillet Type:
Variable: **Linear**
First edge: **4**
Second edge: **0**
OK   ]

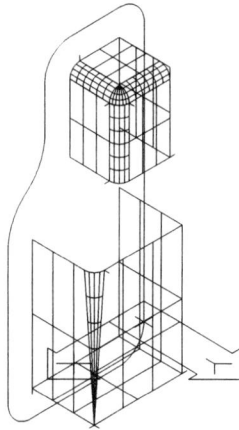

Figure 3.31 Three intersecting fillets rounded off and a variable fillet formed

To proceed making the rear part of the camera, you should change the viewing direction of the display. Run the VPOINT command. See Figure 3.32.

<View>   **<3D Viewpoint>**   **<Vector>**

Command: **VPOINT**
Rotate/<View point>: **R**
Enter angle in XY plane from X axis: **135**
Enter angle from XY plane: **35**

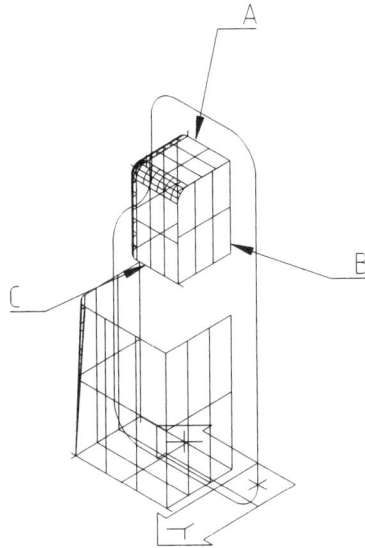

Figure 3.32 New viewing position

Like you did for the other corner you created, use the AMFILLETSF command to create two fillets with the Base surface option. See Figure 3.33.

<Surfaces>   **<Create Surface>**   **<Fillet...>**

Command: **AMFILLETSF**
Select first surface: **[Select A (Figure 3.32.]**
Select second surface: **[Select B (Figure 3.32).]**

[Trim:
First surface: **Yes**
Second surface: **Yes**
Fillet Type:
Base surface: **Yes**
Fillet Radius: **4**
OK   ]

<Surfaces>   **<Create Surface>**   **<Fillet...>**

Command: **AMFILLETSF**
Select first surface: **[Select B (Figure 3.32.]**
Select second surface: **[Select C (Figure 3.32).]**

[Trim:
First surface: **Yes**
Second surface: **Yes**
Fillet Type:
Base surface: **Yes**
Fillet Radius: **4**
OK   ]

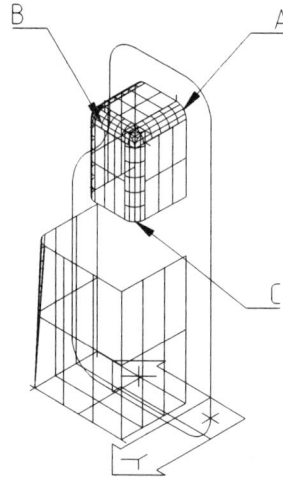

Figure 3.33 Intersecting fillets formed

Because you have three intersecting fillets meeting at a corner, run the AMCORNER command. See Figure 3.34.

\<Surfaces\>   **\<Create Surface\>**   **\<Corner Fillet\>**

Command: **AMCORNER**
(Trim = Yes)
Trim/\<Select first fillet surface\>: **[Select A (Figure 3.33).]**
Trim/\<Select second fillet surface\>: **[Select B (Figure 3.33).]**
Trim/\<Select third fillet surface\>: **[Select C (Figure 3.33).]**

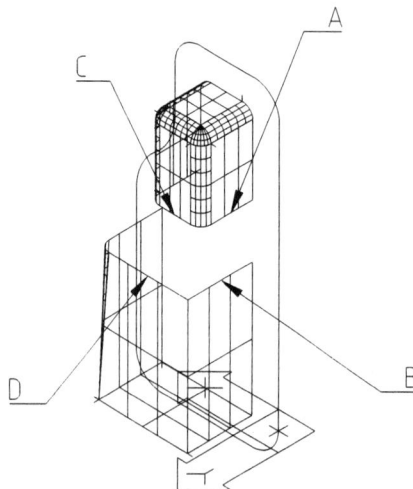

Figure 3.34 Three intersecting fillets rounded off

Fillet surfaces are used to round the edges between touching or intersecting surfaces. With two surfaces are situated apart, you can join them together by blending. Run the AMBLEND command. See Figure 3.35.

<Surfaces> **<Create Surface>** **<Blend>**

Command: **AMBLEND**
Select first wire: **[Select A (Figure 3.34).]**
Select second wire: **[Select B (Figure 3.34).]**
Weights/<Select third wire>: **[Enter]**

<Surfaces> **<Create Surface>** **<Blend>**

Command: **AMBLEND**
Select first wire: **[Select C (Figure 3.34).]**
Select second wire: **[Select D (Figure 3.34).]**
Weights/<Select third wire>: **[Enter]**

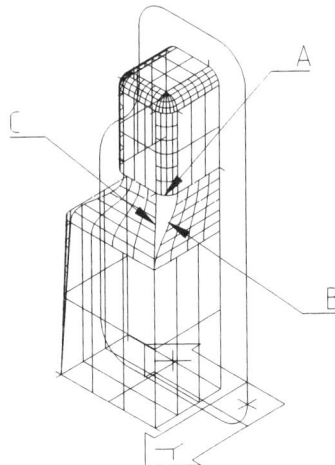

Figure 3.35 Two blended surfaces created

To blend two surfaces, their normal directions must be consistent. If not, the result will be very unpredictable. If your blended surface does not resemble Figure 3.35, you should erase the blended surface, run the AMEDITSF command to reverse the normal direction of one of the original surfaces, and then use the AMBLEND command again.

In addition to blending two surfaces, you can blend three surfaces. Run the AMBLEND command again. See Figure 3.36.

<Surfaces> **<Create Surface>** **<Blend>**

Command: **AMBLEND**
Select first wire: **[Select A (Figure 3.35).]**
Select second wire: **[Select B (Figure 3.35).]**
Weights/<Select third wire>: **[Select C (Figure 3.35).]**
Select fourth wire: **[Enter]**

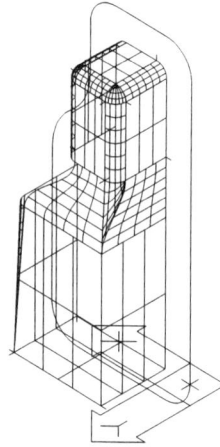

Figure 3.36 Three edges blended

This side of the rear part of the camera is complete. Change the viewing position again by running the VPOINT command. See Figure 3.37.

\<View\>   **\<3D Viewpoint\>**   **\<Vector\>**

Command: **VPOINT**
Rotate/\<View point\>: **R**
Enter angle in˙XY plane from X axis: **45**
Enter angle from XY plane: **35**

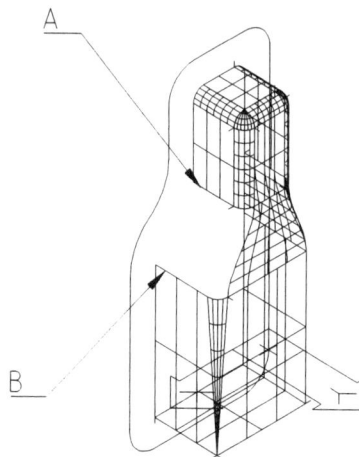

Figure 3.37 Previous viewing position

Apply the AMBLEND command on two surfaces. See Figure 3.38.

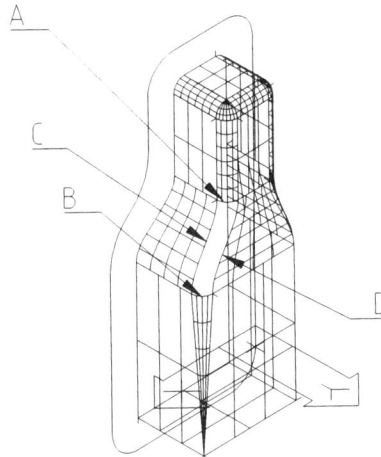

Figure 3.38 Two edges blended

<Surfaces>   <Create Surface>   <Blend>

Command: **AMBLEND**
Select first wire: **[Select A (Figure 3.37).]**
Select second wire: **[Select B (Figure 3.37).]**
Weights/<Select third wire>: **[Enter]**

The AMBLEND command can work on two, three, or four edges. Run this command to blend four edges. See Figure 3.39.

<Surfaces>   <Create Surface>   <Blend>

Command: **AMBLEND**
Select first wire: **[Select A (Figure 3.38).]**
Select second wire: **[Select B (Figure 3.38).]**
Weights/<Select third wire>: **[Select C (Figure 3.38).]**
Select fourth wire: **[Select D (Figure 3.38).]**

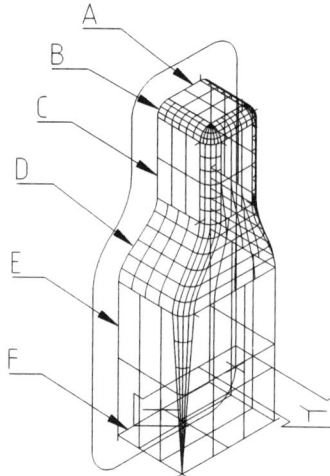

Figure 3.39 Four edges blended

The rear part of the camera needs one more surface — a trimmed planar surface. Figure 3.43 shows what is required.

To produce this surface, you have to make a trimmed planar surface that is based on the outer wireframe, entity W21, and then cut an opening on it.

To cut the opening, you need some wireframes to work on. To make these wireframes, you will copy them from the edges of the surfaces that you have just created.

Create a new layer called FLOW with color blue, and set it as the current layer.

<Data>   <Layers...>

Command: **DDLMODES**

Layer	Color
FLOW	**blue**

Current layer: **FLOW**

Use the AMEDGE command to copy six edges. Watch the elicitation of six blue wireframes on your screen. You may not readily perceive them because they reside on existing edges.

<Surfaces>   <Edit Trim Edges>   <Copy Edge>

Command: **AMEDGE**
(Output = Polyline)
Copy edge/Output/Show nodes/Untrim/<Extract loop>: **COPY**
Select surface edge: **[Select A, B, C, D, E, and F (Figure 3.39).]**
Select surface edge: **[Enter]**

After copying six edges, change to another viewing position by using the VPOINT command. See Figure 3.40.

<View>   **<3D Viewpoint>**   **<Vector>**

Command: **VPOINT**
Rotate/<View point>: **R**
Enter angle in XY plane from X axis: **135**
Enter angle from XY plane: **35**

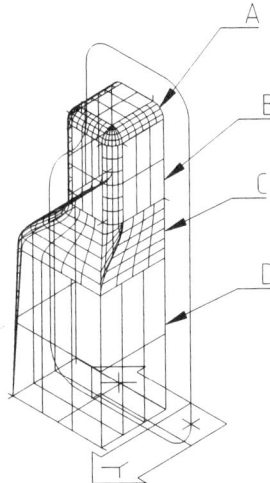

Figure 3.40 New viewing position

Use the AMEDGE command to copy four more edges. Again, watch for four blue wireframes to appear on existing edges.

<Surfaces>   **<Edit Trim Edges>**   **<Copy Edge>**

Command: **AMEDGE**
(Output = Polyline)
Copy edge/Output/Show nodes/Untrim/<Extract loop>: **COPY**
Select surface edge: **[Select A, B, C, and D (Figure 3.40).]**
Select surface edge: **[Enter]**

In order to see the copied edges clearly, turn off the layer SURF2. See Figure 3.41.

<Data>   **<Layers...>**

Command: **DDLMODES**

Layer	
SURF2	**Off**

Current layer: **FLOW**

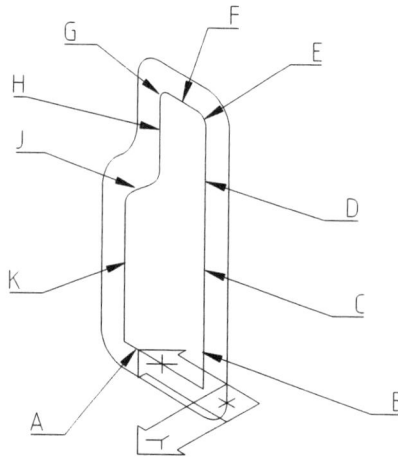

Figure 3.41 Layer SURF2 turned off

The copied edges are separate entities. Join them together by using the AMJOIN3D command so that you can handle them collectively as one wireframe.

<Surfaces>   <Edit Wireframe>   <Join...>

Command: **AMJOIN3D**

> [Mode: **Automatic**
> Output: **3D Polyline**
> **OK**   ]

Select start wire: **[Select A (Figure 3.41).]**
Select wires to join: **[Select B, C, D, E, F, G, H, J, and K (Figure 3.41).]**
Select wires to join: **[Enter]**
Reverse? Yes/<No>: **NO**

Return to the wireframe W21 again. Based on this 3D polyline, run the AMPLANE command to produce a trimmed planar surface. See Figure 3.42.

Before making the planar surface, run the AMSURFVARS command to set the polyline fit length to 10 units. Doing so will set the system variable AMPFITLEN to 10. Thus, any segment with a length greater than 10 units will remain as a flat segment.

<Surfaces>   <Preferences...>

Command: **AMSURFVARS**

> [Surface/Spline Options:
> Polyline fit Length: **10**
> **OK**   ]

<Surfaces>   <Create Surface>   <Planar Trim>

Command: **AMPLANE**
Plane/Wires/<First corner>: **W**

Select wires: **GROUP**
Enter group name: **W21**
Select wires: **[Enter]**

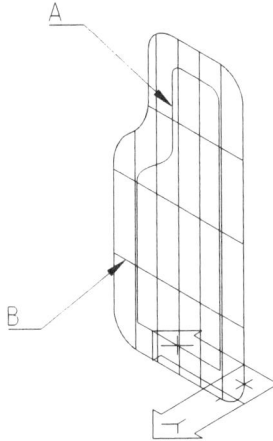

Figure 3.42 Trimmed planar surface created

To cut an opening on the trimmed planar surface, run the AMPROJECT command. Use the polyline joined from the copied edges as the wireframe to project. See Figure 3.43.

<Surfaces>   <Edit Surface>   <Project Trim...>

Command: **AMPROJECT**
Select wires to project: **[Select A (Figure 3.42).]**
Select wires to project: **[Enter]**
Select target surfaces: **[Select B (Figure 3.42).]**
Select target surfaces: **[Enter]**

    [Direction: **Normal to surface**
     Output type: **Trim surface**
     OK   ]

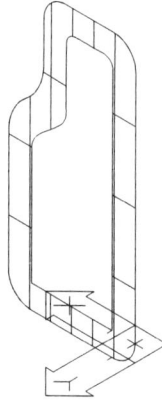

Figure 3.43 Planar surface trimmed

The trimmed planar surface now resides on the FLOW layer. Move it to layer SURF2. Then, turn on layer SURF2 and turn off layer FLOW.

Edit> **<Properties...>**

Select objects: **[Select the trimmed planar surface.]**
Select objects: **[Enter]**

[Properties
Layer... **SURF2**
OK]

<Data> **<Layers...>**

Command: **DDLMODES**

Layer	
SURF2	**On**
FLOW	**Off**

Current layer: **SURF2**

You have completed the rear part of the video camera. Refer back to Figure 3.24.

3.3 Body of the Camera

Figure 3.44 shows the completed surface model for the body of the video camera. The body consists of four parts — the bottom, right side, top, and left side. First, you will create the bottom part, which is a ruled surface. Then, you will create a lofted U surface to form the right side of the body. Next, you will create a swept surface for the top surface of the body. The fourth part of the model, the left side of the body, is a bit more complicated. You will make a surface of revolution and a surface of extrusion. Then, you will form a fillet surface between them to complete the model.

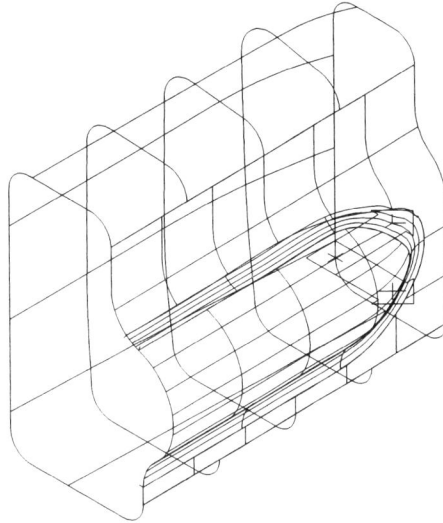

Figure 3.44 Surfaces for the body of the video camera

Turn off layer SURF2, and create a new layer called SURF3 with color magenta. Set this layer as the current layer. You will create the body portion of the camera on this layer.

<Data>   <Layers...>

Command: **DDLMODES**

Layer	Color	
SURF2		**Off**
SURF3	**magenta**	

Current layer: **SURF3**

Use the accelerator key [8] to return to an isometric view. Apply the AMVISIBLE command to unhide the entity group W31. See Figure 3.45.

Command: **8**

<Surfaces>   <Object Visibility...>

Command: **AMVISIBLE**

      [Unhide
       Select   ]

Select objects to unhide: **GROUP**
Enter group name: **W31**
Select objects to unhide: **[Enter]**

    [OK]

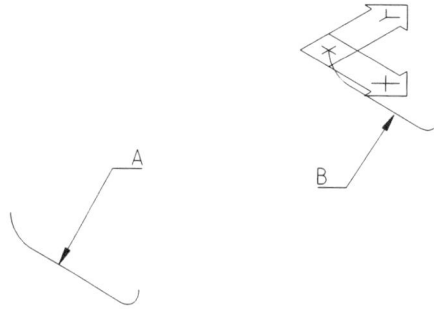

Figure 3.45 Entity group W31 unhidden

You will use the entity group W31 to create a ruled surface. A ruled surface is a surface of interpolation between two boundary wireframes. Entity group W31 is a set of polylines. During surface creation, AutoSurf converts polylines to splines according to the variable AMPFITLEN setting.

Apply the AMRULE command. See Figure 3.46.

<Surfaces>   <Create Surface>   <Rule>

Command: **AMRULE**
Select first wire: **[Select A (Figure 3.45.]**
Select second wire: **[Select B (Figure 3.45).]**

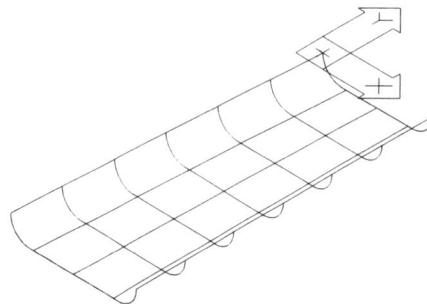

Figure 3.46 Ruled surface created

Run the AMVISIBLE command to unhide the entity group W32. Then, apply the AMLOFTU command on entity group W32 to produce a lofted U surface. A lofted U surface is a surface of interpolation between a series of cross sections. See Figure 3.47.

<Surfaces>   <Object Visibility...>

Command: **AMVISIBLE**

**[Unhide
Select   ]**

Select objects to unhide: **GROUP**
Enter group name: **W32**
Select objects to unhide: **[Enter]**

　　　[OK]

<Surfaces>   <Create Surface>   <Loft U...>

Command: **AMLOFTU**
Select U wires: **GROUP**
Enter group name: **W32**
Select U wires: **[Enter]**

　　　[Input Wires
　　　Curve Direction Align   **Yes**
　　　Curve Fit   Smooth   **Yes**
　　　OK   ]

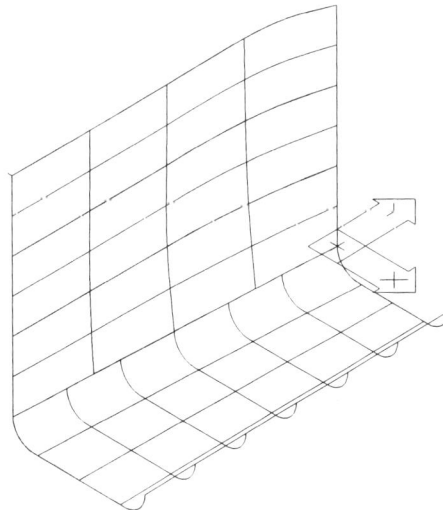

Figure 3.47 Lofted U surface created

The third surface for the body is a swept surface. You can create a swept surface by moving one or more profiles along one or two rails.

Run the AMVISIBLE command to unhide the entity groups W33A and W33B. See Figure 3.48.

<Surfaces>   <Object Visibility...>

Command: **AMVISIBLE**

　　　[**Unhide**
　　　Select   ]

Select objects to unhide: **GROUP**
Enter group name: **W33A,W33B**

Select objects to unhide: **[Enter]**

    [OK]

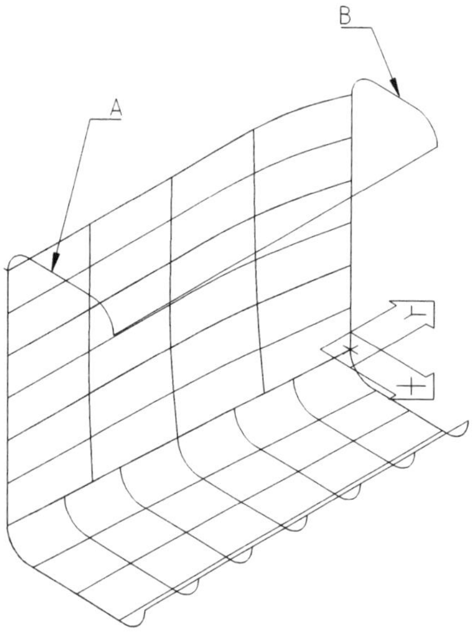

Figure 3.48 Entity groups W33A and W33B unhidden

Using entity W33A as the cross sections and entity W33B as the rails, generate a swept surface by using the AMSWEEPSF command. See Figure 3.49.

    <Surfaces>   **<Create Surface>**   **<Sweep>**

    Command: **AMSWEEPSF**
    Select cross sections: **GROUP**
    Enter group name: **W33A**
    Select cross sections: **[Enter]**
    Select rails: **[Select A (Figure 3.48).]**
    Select rails: **[Select B (Figure 3.48).]**

        [Transition   **Scale**
          OK   ]

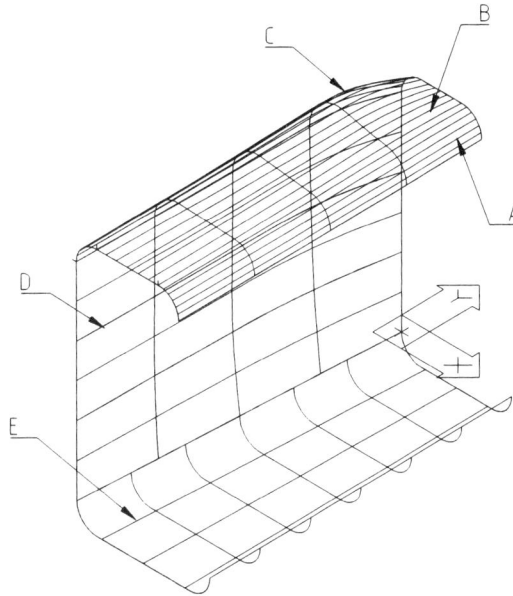

Figure 3.49 Swept surface created

Do you recall that you can join untrimmed surfaces? Run the AMJOINSF command to join the ruled surface, the lofted U surface, and the swept surface. Then, put the joined surface in an entity group called S31 by using the GROUP command. See Figure 3.50.

\<Surfaces\>   \<Edit Surface\>   \<Join\>

Command: **AMJOINSF**
Select surfaces to join: **[Select A, B, C, D, and E (Figure 3.49).]**
Select surfaces to join: **[Enter]**

[Standard Toolbar]   [Object Group]

Command: **GROUP**

[Group name: **S31**
  Selectable: **Yes**
  New:]

Select objects for grouping:
Select objects: **LAST**
Select objects: **[Enter]**

    [OK]

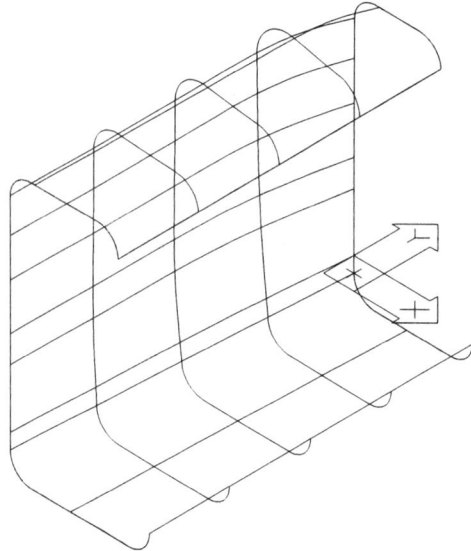

Figure 3.50 Ruled surface, lofted U surface, and swept surface joined

Now, the three surfaces are joined and are a single surface. Compare Figure 3.50 with Figure 3.51 to see the difference.

Use the AMVISIBLE command to hide all the entities, and then unhide the entity groups W34A, W34B, and W34C. You will use these entities to build the last surface of the body of the camera. See Figure 3.51.

\<Surfaces\>   \<Object Visibility...\>

Command: **AMVISIBLE**

> **[Hide**
> **All**
> **OK**   ]

\<Surfaces\>   \<Object Visibility...\>

Command: **AMVISIBLE**

> **[Unhide**
> **Select**   ]

Select objects to unhide: **GROUP**
Enter group name: **W34A,W34B,W34C**
Select objects to unhide: **[Enter]**

> **[OK]**

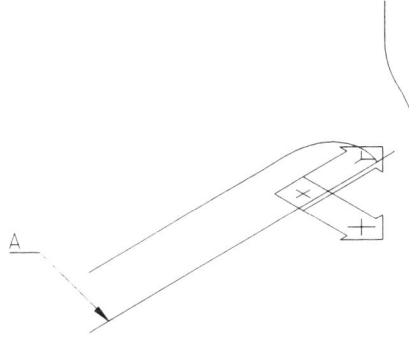

Figure 3.51 Entity groups W34A, W34B, and W34C

You will use the entity groups W34A and W34B to create a surface of revolution, and use the entity group W34C to create a surface of extrusion.

Run the AMREVOLVESF command on the entity groups W34A and W34B. See Figure 3.52.

<Surfaces>   <Create Surface>   <Revolve>

Command: **AMREVOLVESF**
Select path curves: **GROUP**
Enter group name: **W34A**
Select path curves: **[Enter]**
Axis of revolution: Wire/<Start point of axis>: **WIRE**
Select wire: **[Select A (Figure 3.51).]**
Start angle: **0**
Included angle (+=ccw, -=cw) <Full circle>: **180**

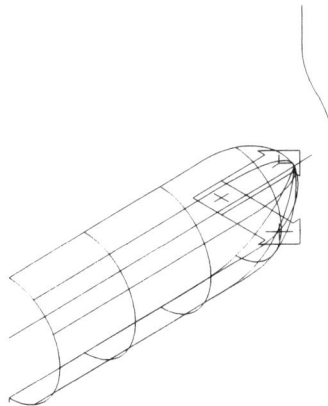

Figure 3.52 Surface of revolution created

Use the AMEXTRUDESF command to form a surface of extrusion from the entity group W34C. See Figure 3.53.

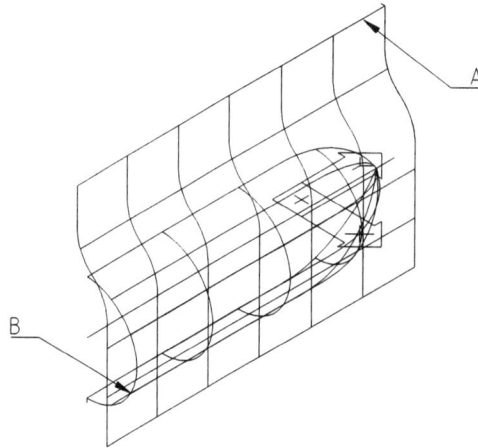

Figure 3.53 Surface of extrusion created

\<Surfaces\>   \<Create Surface\>   \<Extrude\>

Command: **AMEXTRUDESF**
Select wires: **GROUP**
Enter group name: **W34C**
Select wires: **[Enter]**
Direction: Viewdir/Wire/X/Y/Z/\<Start point\>: **Y**
Distance: **-150**
Flip/\<Accept\>: **ACCEPT**
Taper angle: **0**

To complete the body of the camera, use the AMFILLETSF command to trim and fillet at the intersection of the surface of extrusion and the surface of revolution. See Figure 3.54.

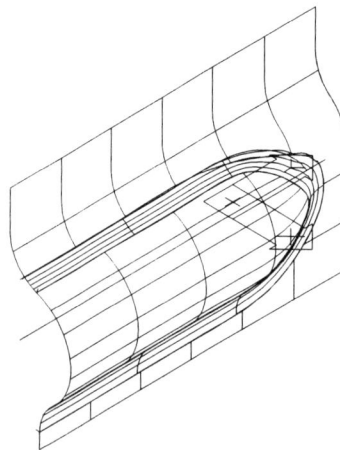

Figure 3.54 Revolved surface and extruded surface filleted and trimmed

<Surfaces>   **<Create Surface>**   **<Fillet...>**

Command: **AMFILLETSF**
Select first surface: **[Select A (Figure 3.53).]**
Select second surface: **[Select B (Figure 3.53).]**

[Trim:
First surface: **Yes**
Second surface: **Yes**
Fillet Radius: **14**
OK   ]

You have completed the body of the camera. Take a look at the entire camera body by unhiding the surface S11 and hiding all the related wireframes. Run the AMVISIBLE command. Compare your work with Figure 3.44.

<Surfaces>   **<Object Visibility...>**

Command: **AMVISIBLE**

[**Unhide**
 Select   ]

Select objects to unhide: **GROUP**
Enter group name: **S31**
Select objects to unhide: **[Enter]**

[**Hide**
 Select   ]

Select objects to hide: **GROUP**
Enter group name: **W***
Select objects to hide: **[Enter]**

[**OK**]

3.4 Eyepiece of the Camera

Figure 3.55 is the surfaces for the eyepiece of the video camera. The eyepiece consists of three surfaces — a primitive cylindrical surface, a tubular surface, and a ruled surface.

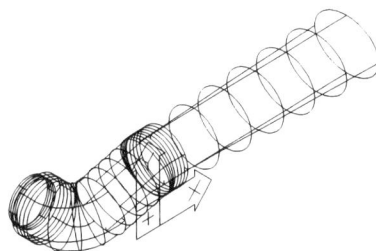

Figure 3.55 Surfaces for the eyepiece of the video camera

Create a new layer named SURF4 with color blue, and make this the current layer.

<Data>   **<Layers...>**

Command: **DDLMODES**

Layer	Color
SURF4	**blue**

Current layer: **SURF4**

Unhide the entity groups W41 and W42. See Figure 3.56.

<Surfaces>   **<Object Visibility...>**

Command: **AMVISIBLE**

       [Unhide
       Select   **]**

Select objects to unhide: **GROUP**
Enter group name: **W41,W42**
Select objects to unhide: **[Enter]**

      [OK]

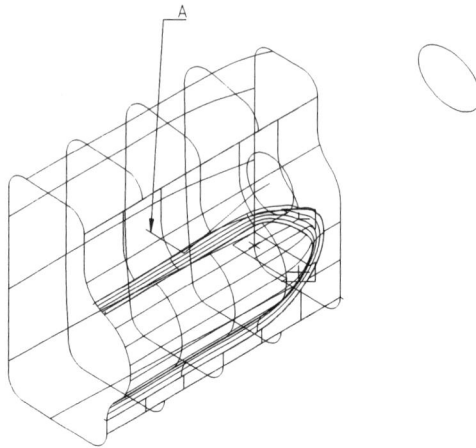

Figure 3.56 Wireframes for the eyepiece unhidden

Align the UCS with the current YZ plane. Then, use the AMPRIMSF command to create a primitive cylindrical surface. The orientation of the cylinder created depends on the current UCS. See Figure 3.57.

[UCS]   **[Z Axis Vector UCS]**

Command: **UCS**
Origin/ZAxis/3point/OBject/View/X/Y/Z/Prev/Restore/Save/Del/?/<World>: **ZA**

Origin point: **0,0,0**
Point on positive portion of Z-axis: **1,0**

<Surfaces>   <Primitives>   <Cylinder>

Command: **AMPRIMSF**
Cone/cYlinder/Torus/<Sphere>: **Y**
Base center point: **END** of **[Select A (Figure 3.56).]**
Diameter/<Radius> of base: **11**
Height: **-15**
Start angle: **0**
Included angle (+=ccw, -=cw) <Full circle>: **[Enter]**

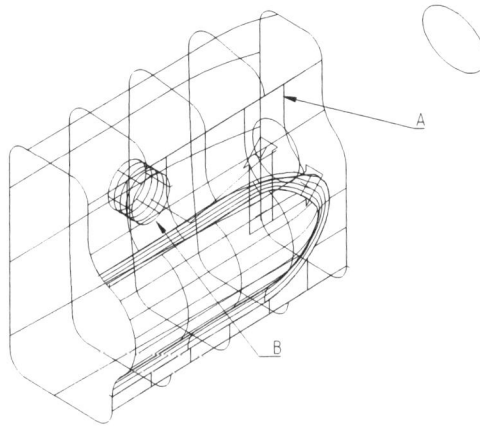

Figure 3.57 Primitive cylindrical surface created

After making the primitive cylindrical surface, use the AMINTERSF command to trim off the excess surface at the intersection of the cylindrical surface and the extruded surface of the body. See Figure 3.58.

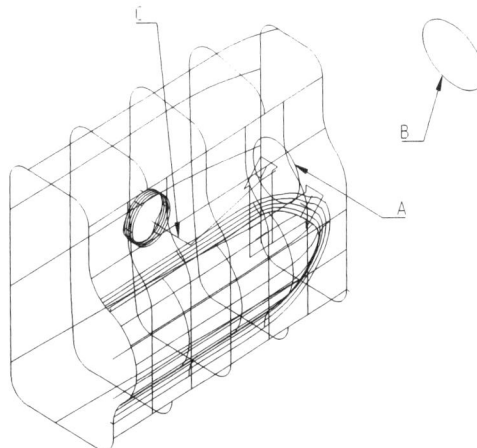

Figure 3.58 Primitive cylindrical surface intersected with the right side of the body

\<Surfaces\>   \<Edit Surface\>   \<Intersect Trim...\>

Command: **AMINTERSF**
Select first surface: **[Select A (Figure 3.57).]**
Select second surface: **[Select B (Figure 3.57).]**

[Trim surface:
 First surface: **Yes**
 Second surface: **Yes**
 OK   ]

To complete the eyepiece, use the AMRULE command to make a ruled surface, and use the AMTUBE command to make a tubular surface. See Figure 3.59.

\<Surfaces\>   \<Create Surface\>   \<Rule\>

Command: **AMRULE**
Select first wire: **[Select A (Figure 3.58).]**
Select second wire: **[Select B (Figure 3.58).]**

\<Surfaces\>   \<Create Surface\>   \<Tubular\>

Command: **AMTUBE**
Select wire: **[Select C (Figure 3.58).]**
Tube diameter: **30**
Manual/\<Automatic\>: **A**
Bend radius for all: **16**

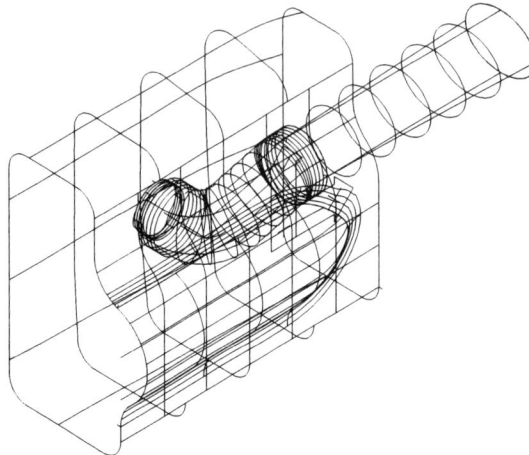

Figure 3.59 Eyepiece of the video camera completed

You have completed the surfaces for the eyepiece of the video camera.

3.5 **Completed Video Camera**

In the previous sections of this chapter, you created four sets of surfaces, which reside on layers SURF1, SURF2, SURF3, and SURF4.

To see the entire surface model, unhide all entities, turn off layers WIRE1 and WIRE2, and turn on layers SURF1 and SURF2.

<Surfaces>   **<Object Visibility...>**

Command: **AMVISIBLE**

> **[Unhide**
> **All**
> **OK**   **]**

<Data>   **<Layers...>**

Command: **DDLMODES**

Layer	·
SURF1	**On**
SURF2	**On**
WIRE1	**Off**
WIRE2	**Off**

Reset the UCS to WORLD.

[UCS]   **[World UCS]**

Command: **UCS**
Origin/ZAxis/3point/Entity/View/X/Y/Z/Prev/Restore/Save/Del/?/<World>: **W**

Finally, use the accelerator key [4] to divide the screen display into four viewports, which show the top view, the front view, the side view, and the isometric view of the camera. See Figure 3.60.

Command: **4**

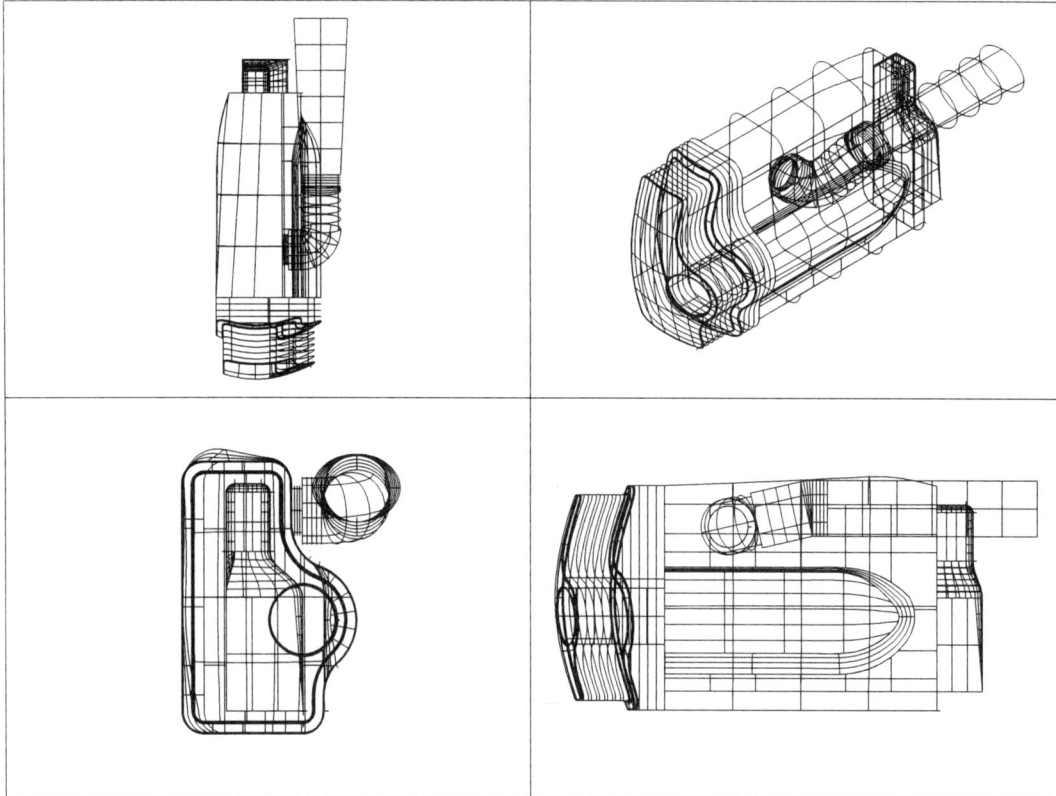

Figure 3.60 Four views of the video camera

3.6   Summary

In this chapter, you have practiced the following AutoSurf commands:

AM2SF	AMBLEND	AMCORNER
AMEXTRUDESF	AMFILLETSF	AMLOFTU
AMLOFTUV	AMOFFSETSF	AMPLANE
AMPRIMSF	AMREVOLVESF	AMRULE
AMSWEEPSF	AMTUBE	AMBREAK
AMEDGE	AMEDITSF	AMINTERSF
AMJOINSF	AMLENGTHEN	AMPROJECT
AMREFINESF		

For a brief explanation of these commands, refer to the appendix of this book.

In making this surface model, you have used most of the surface creating and editing commands of AutoSurf. By now, you should be able to create extruded surfaces, revolved surfaces, swept surfaces, tubular surfaces, lofted U surfaces, lofted UV surfaces, planar surfaces, ruled surfaces, blended surfaces, corner fillet surfaces, fillet surfaces, and offset surfaces. You should also be able to convert an AutoCAD object into a set of surfaces. In addition to surface creation, you should be able to break surfaces, intersect surfaces, join surfaces, lengthen surfaces, project and trim surfaces, refine surfaces, and flip the normal of a surface.

In the next chapter, you will practice creating another model and learn to use other utility commands.

3.7 Exercise

To further enhance your knowledge of surface modeling, you will make the NURBS surface model of a scale model car based on the wireframes you created in Chapter 2.

Open the file CAR.DWG that you saved in Chapter 2, and complete the surface model of the car, as shown in Figure 3.61.

Figure 3.61 Completed surface model

Because you learned how to use the surface creation and editing commands in this chapter, this section will provide only a brief outline of the steps you need to follow to complete the surface model of the car.

To begin, check your drawing to see if you have organized the entities into four entity groups: BD, TP, GH, and WH. If not, go back to Chapter 2 to finish the grouping.

Hide all the entity groups except BD. Make this group not selectable, so that you can select each member of it individually. See Figure 3.62.

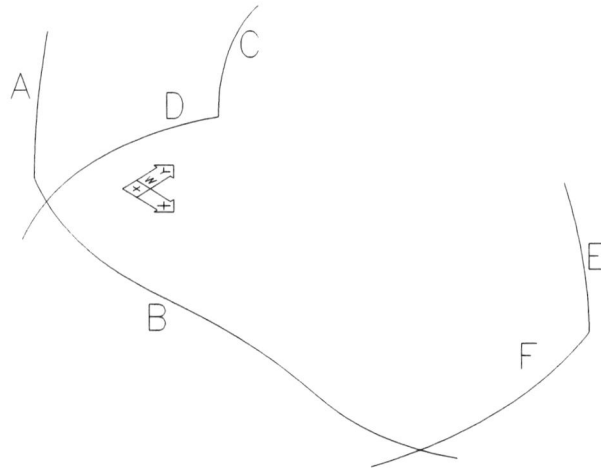

Figure 3.62 Entity group BD

Create four swept surfaces with a parallel orientation of the cross section. The first surface uses A (Figure 3.62) as the cross section and B (Figure 3.62) as the rail. The second surface uses C (Figure 3.62) as the cross section and D (Figure 3.62) as the rail. The third surface uses E (Figure 3.62) as the cross section and F (Figure 3.62) as the rail. The fourth surface is a mirror of first surface about the current X axis.

After producing the swept surfaces, make the group BD selectable again. Then, hide the group BD and unhide the group TP. Make TP unselectable. See Figure 3.63.

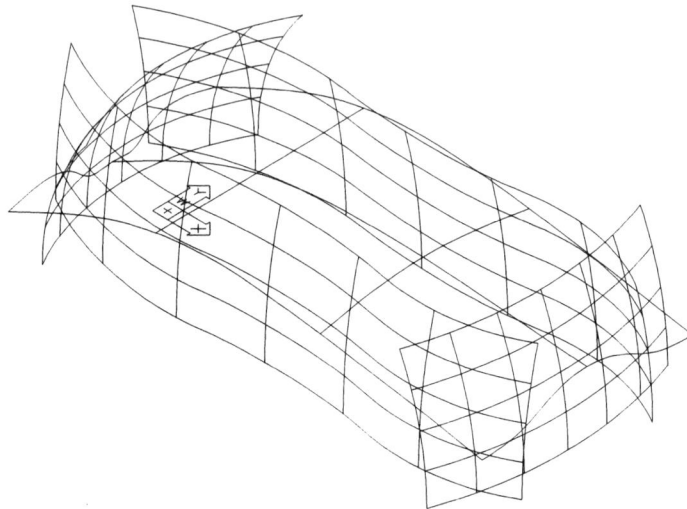

Figure 3.63 Four swept surfaces created and the entity group TP unhidden

Using the seven splines of group TP as UV lines, create a lofted UV surface. Then, make the group TP selectable again, and hide it. See Figure 2.64.

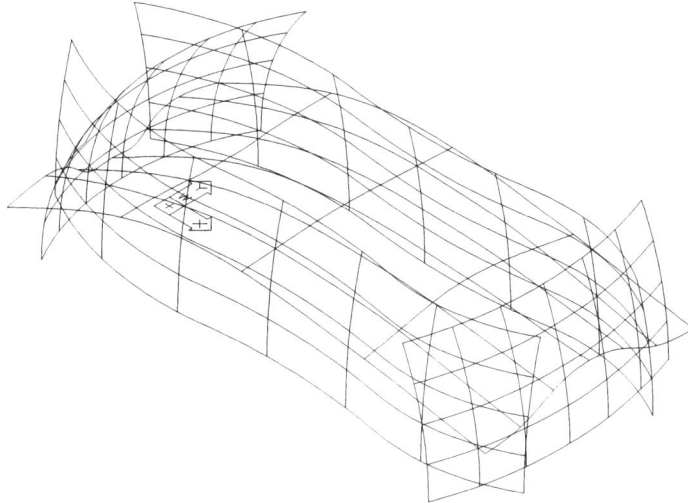

Figure 3.64 Lofted UV surface created

As shown in Figure 3.65, create two fillet surfaces of radius 10 units. In doing the fillets, trim both objects and use the extended fillet type.

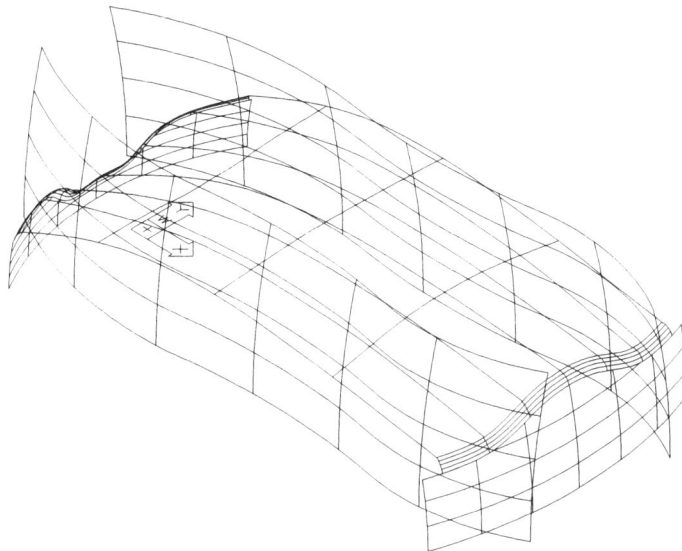

Figure 3.65 Two extended fillet surfaces created

As shown in Figure 3.66, create six more fillet surfaces of constant radius 10 units. Trim both surfaces. The fillet type is extended and base surface.

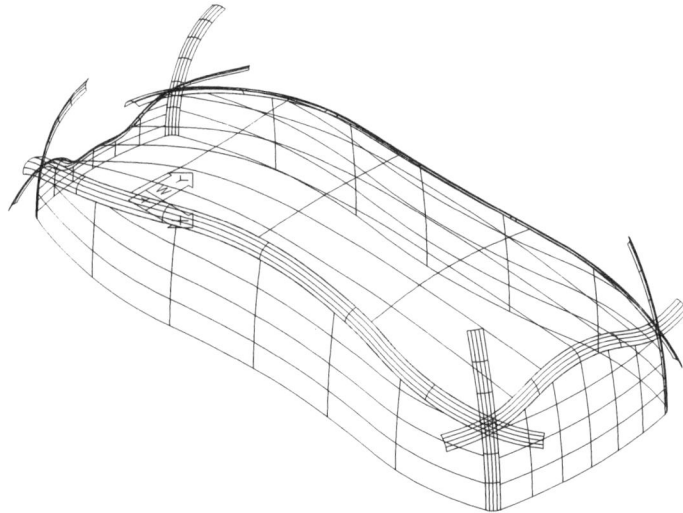

Figure 3.66 Four extended base surface fillets created

Round the four corners. Then, unhide the entity group WH and make it unselectable. See Figure 3.67.

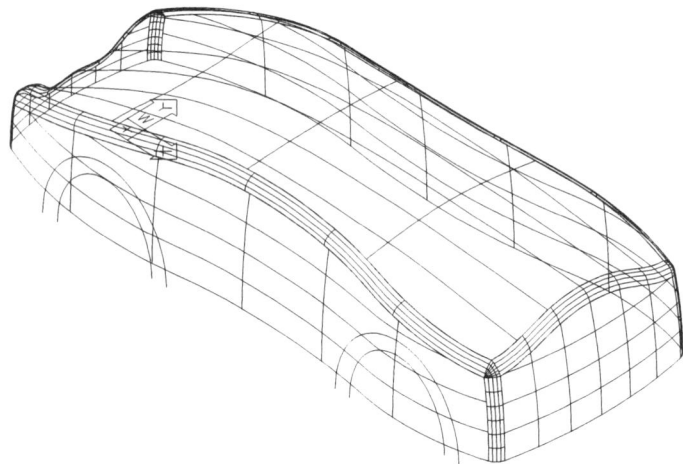

Figure 3.67 Four corners rounded and group WH unhidden

Project the two half ellipses in the Y direction of WCS onto the vertical surface of the car model. See Figure 3.68.

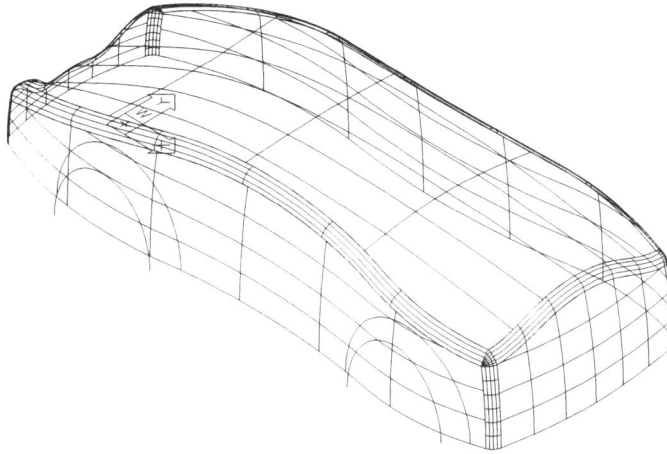

Figure 3.68 Two wireframes projected onto a surface

Use the shortcut key [6] to set to a front view as shown in Figure 3.69. Then, 3D offset the two projected wireframes for a distance of 4 units toward the center of the half ellipses. Remember, the AMOFFSET3D command is view dependent. You must switch to a proper view before running this command.

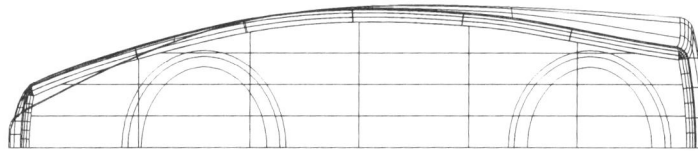

Figure 3.69 View set to front and projected wireframes offset

Switch back to an isometric view. Then, translate the two newly created offset wireframes a distance of 1 unit towards the negative Y direction of WCS. See Figure 3.70.

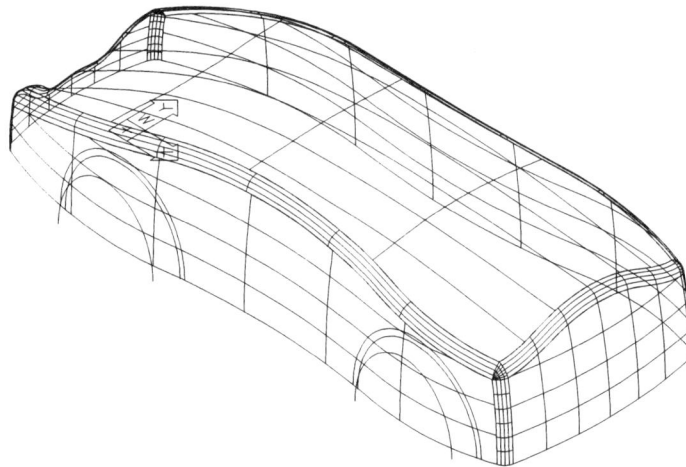

Figure 3.70 Two offset wireframes translated

Create two lofted U surfaces, using the projected wireframes, the offset wireframes, and the two half ellipses. In making the lofted U surfaces, align and smooth the input wireframes and respace the curve ends.

After making the lofted U surface, use the projected wireframes to trim the vertical surface of the car body. See Figure 3.71.

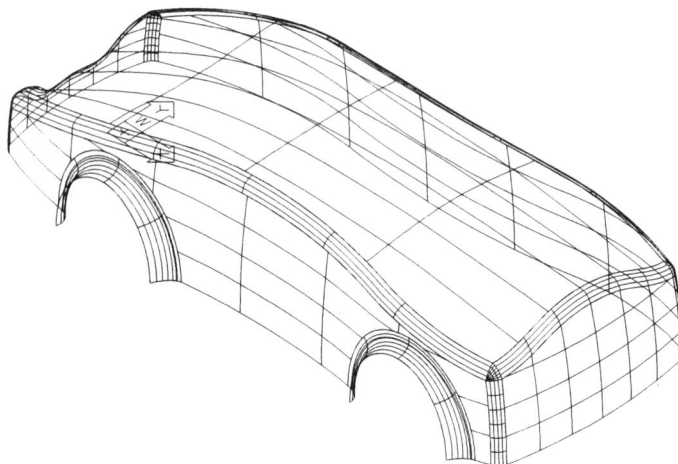

Figure 3.71 Two lofted U surfaces created and the vertical surface trimmed

As shown in Figure 3.72, rotate the UCS above the X axis for 90°. Then, make two planar surfaces. The size of them is unimportant as long as they are larger than the wheel openings.

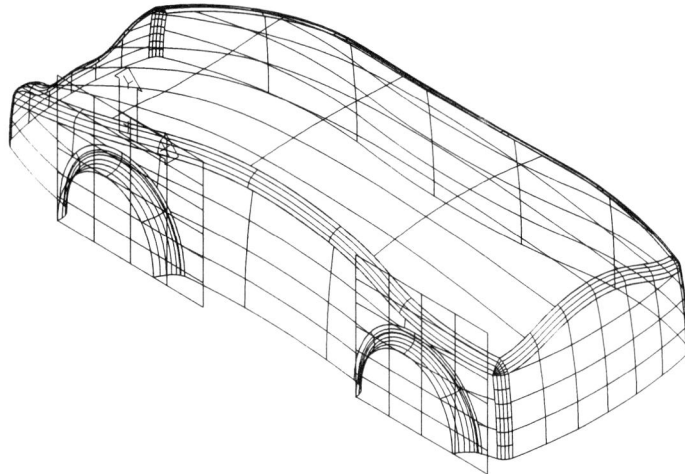

Figure 3.72 UCS rotated and two planar surfaces created

Form fillet surfaces of radius 5 units between the planar surfaces and the lofted U surfaces. Then, erase the planar surfaces. See Figure 3.73.

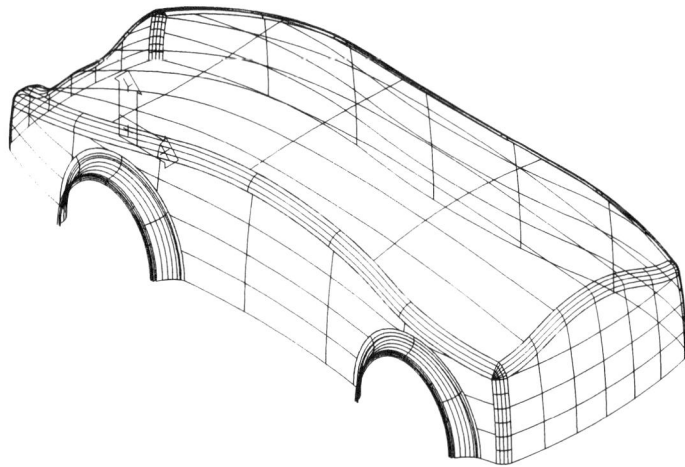

Figure 3.73 Fillet surfaces created and planar surfaces erased

The model car is symmetrical. To complete the other side of the car, you can erase the vertical surface of the other side, and mirror the finished side. Unhide the entity group GH, and make it unselectable. See Figure 3.74.

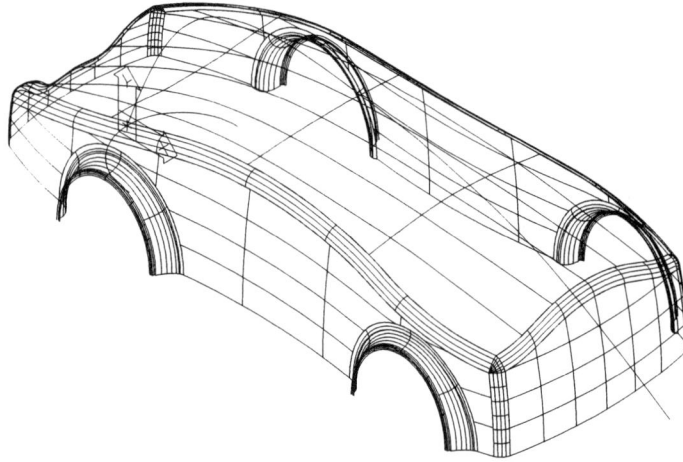

Figure 3.74 Other side of the car completed

Using the entity group GH, create a sweep surface with normal orientation. See Figure 3.75.

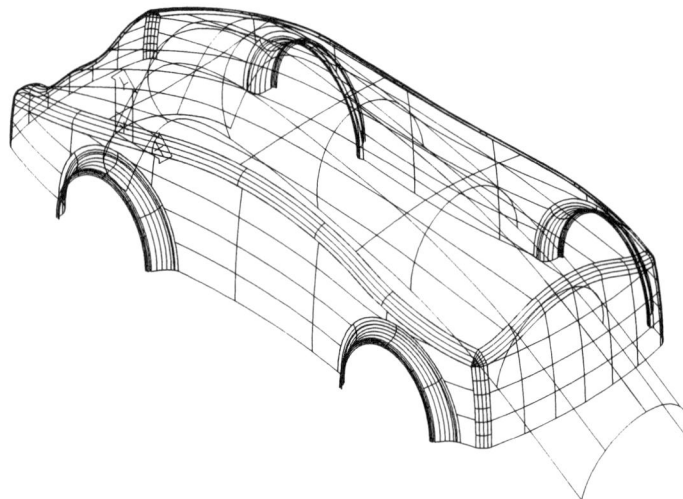

Figure 3.75 Swept surface created

Form a fillet surface of radius 3 units between the newly created swept surface and the top surface of the car. Then, turn off the layer WIRE. The surface model is completed. See Figure 3.76.

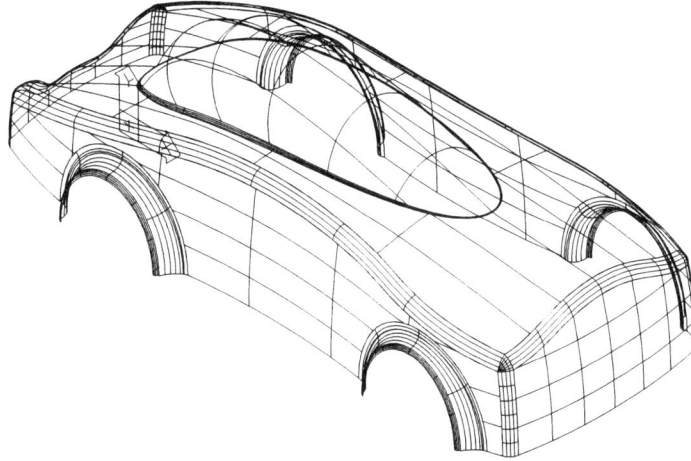

Figure 3.76 Fillet surface formed between the swept surface and the top surface

Save your drawing.

Chapter 4
Additional Practice on Model Creation

The design of the video camera model produced in the last two chapters enabled you to practice making all kinds of AutoSurf surfaces. You applied all the AutoSurf surface creation and editing commands along with AutoSurf wireframes, display controls, and utility commands. When you employ AutoSurf to build your own surface model, you do not necessarily have to use all kinds of surfaces. Instead, you should choose the most appropriate types of surface that meet your design needs.

To start, you must create a series of free-form surfaces to construct the main body of the model. The surface can be a planar surface, extruded surface, revolved surface, ruled surface, swept surface, lofted U surface, or lofted UV surface. Then, you have to treat the joints between the surfaces by using derived surfaces, such as fillet surfaces, corner surfaces, and blended surfaces. To add features to your design, you can use offset surfaces.

In this chapter, you will work on the surface model of a mobile phone (shown in Figure 4.1) to enhance your knowledge of AutoSurf, and to practice other AutoSurf utility commands. Basically, the process of creating this model is very similar to that of the video camera. You first produce all the necessary wireframes to define the surface contours, and then you create surfaces from the wireframes. In addition to model creation, you will learn how to create flow lines, parting lines, section lines, and augmented lines. Also, you will learn how to compute mass properties from a surface model, and to output a surface model to other applications.

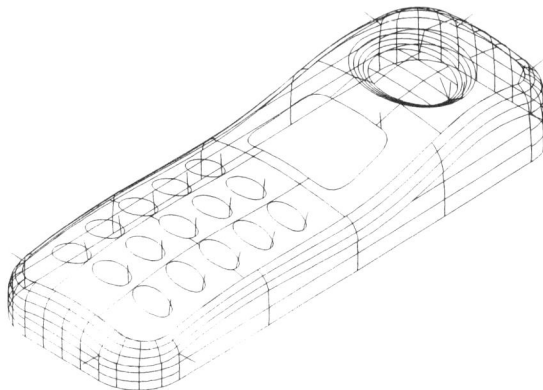

Figure 4.1 NURBS surface model of the mobile phone

131

4.1   Wireframe Creation

The wireframes for this model are very simple. Figure 4.2 shows all the wireframes required for the mobile phone.

The main body of the mobile phone as shown in Figure 4.1 consists of eight side surfaces and a top surface, which are swept surfaces. Each of these surfaces requires a cross section and a rail. For the side surfaces, the three arcs (Figure 4.2) that reside on planes perpendicular to the WCS will be used as the cross sections, and the outer rectangle with four fillet corners will be used as the rails. For the top surface, the spline will be used as the rail, and the arc on the ZX plane will be used as the cross section. The two ellipses will be used to make the earpiece. The trapezoidal object will be used to make the display opening. The suggested button openings are shown in Figure 4.32.

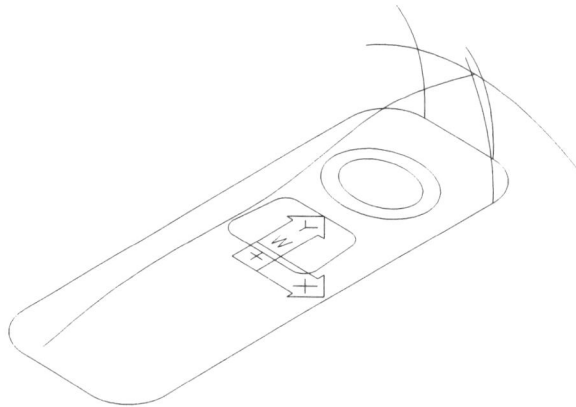

Figure 4.2 Wireframes for the mobile phone

Run the AMSURFVARS command to set the AutoSurf system variables. Select the box Model Size, and specify a size of 150 millimeters. In the Surface Properties column, set U display wires to 3, V display wires to 2, and Vector length to 5 units.

<Surfaces>   <Preferences...>

Command: **AMSURFVARS**

```
[Surface Properties:
U Display Wires:      3
V Display Wires:      2
Vector Length:       5
          [Model Size:
               150    Millimeters ]
OK                                  ]
```

Create a new layer called WIRE with color cyan, and set it as the current layer.

<Data>   <Layers...>

Command: **DDLMODES**

Layer	Color
WIRE	**cyan**

Current layer: **WIRE**

For the sake of clarity, always turn on the UCS icon and set it to the origin position. Run the UCSICON command.

<Options>   **<UCS>**   **<Icon Origin>**

Command: **UCSICON**
ON/OFF/All/Noorigin/ORigin <ON>: **OR**

Use the LINE command to create four line segments.

[Draw]   **[Line]**

Command: **LINE**
From point: **23,65**
To point: **@46<180**
To point: **@130<270**
To point: **@46<0**
To point: **C**

Fit all the entities to the current screen display using the accelerator key [F].

Command: **F**

Make two ellipses with the ELLIPSE command. Also draw a polyline using the PLINE command. See Figure 4.3.

[Draw]   **[Ellipse Center]**

Command: **ELLIPSE**
Arc/Center/<Axis endpoint 1>: **C**
Center of ellipse: **0,40**
Axis endpoint: **@15<0**
<Other axis distance>/Rotation: **12**

[Draw]   **[Ellipse Center]**

Command: **ELLIPSE**
Arc/Center/<Axis endpoint 1>: **C**
Center of ellipse: **0,40**
Axis endpoint: **@11<0**
<Other axis distance>/Rotation: **8**

[Draw]   **[Polyline]**

Command: **PLINE**

From point: **12,22**
Current line-width is 0.0000
Arc/Close/Halfwidth/Length/Undo/Width/<Endpoint of line>: **@24<180**
Arc/Close/Halfwidth/Length/Undo/Width/<Endpoint of line>: **@-2,-20**
Arc/Close/Halfwidth/Length/Undo/Width/<Endpoint of line>: **@28<0**
Arc/Close/Halfwidth/Length/Undo/Width/<Endpoint of line>: **C**

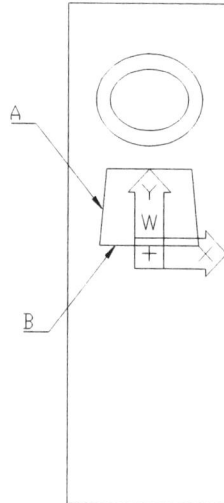

Figure 4.3 Lines, ellipses and polyline created

Execute the FILLET command. Set the fillet radius to 4 units, and then round the corners of the polyline.

    [Modify]   **[Fillet]**

    Command: **FILLET**
    (TRIM mode) Current fillet radius = 0.0000
    Polyline/Radius/Trim/<Select first object>: **R**
    Enter fillet radius: **4**

    [Modify]   **[Fillet]**

    Command: **FILLET**
    (TRIM mode) Current fillet radius = 4.0000
    Polyline/Radius/Trim/<Select first object>: **[Select A (Figure 4.3).]**
    Select second object: **[Select B (Figure 4.3).]**

Repeat the FILLET command three more times on the other corners of the polyline. See Figure 4.4.

    [Modify]   **[Fillet]**

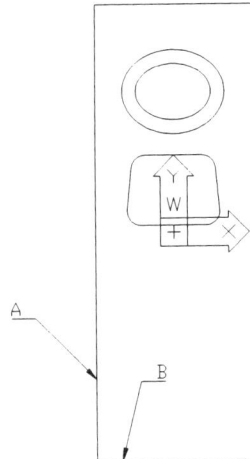

Figure 4.4 Four corners of the polyline filleted

Apply the FILLET command again to set fillet radius to 12 units, and to round the corners of the four lines.

[Modify]   **[Fillet]**

Command: **FILLET**
(TRIM mode) Current fillet radius = 4.0000
Polyline/Radius/Trim/<Select first object>: **R**
Enter fillet radius: **12**

[Modify]   **[Fillet]**

Command: **FILLET**
(TRIM mode) Current fillet radius = 12.0000
Polyline/Radius/Trim/<Select first object>: **[Select A (Figure 4.4).]**
Select second object: **[Select B (Figure 4.4).]**

Repeat the FILLET command three more times. See Figure 4.5.

[Modify]   **[Fillet]**

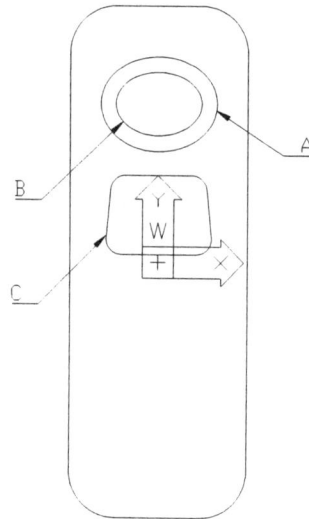

Figure 4.5 Four outer corners filleted

Run the GROUP command. Put the large ellipse in an entity group W1, the small ellipse in W2, and the polyline in W3.

[Standard Toolbar]   **[Object Group]**

Command: **GROUP**

　　　　　[Group name:　　**W1**
　　　　　　Selectable:　　**Yes**
　　　　　　New:　　　　　]

Select objects for grouping:
Select objects: **[Select A (Figure 4.5).]**
Select objects: **[Enter]**

　　　　　[Group name:　　**W2**
　　　　　　Selectable:　　**Yes**
　　　　　　New:　　　　　]

Select objects for grouping:
Select objects: **[Select B (Figure 4.5).]**
Select objects: **[Enter]**

　　　　　[Group name:　　**W3**
　　　　　　Selectable:　　**Yes**
　　　　　　New:　　　　　]

Select objects for grouping:
Select objects: **[Select C (Figure 4.5).]**
Select objects: **[Enter]**

　　　[OK]

Using the accelerator key [8], switch to an isometric view.

    Command: **8**

Set the Z direction of the current UCS to align with the current X axis by using the UCS command.

    [UCS]   **[Z Axis Vector UCS]**

    Command: **UCS**
    Origin/ZAxis/3point/OBject/View/X/Y/Z/Prev/Restore/Save/Del/?/<World>: **ZA**
    Origin point: **0,0,0**
    Point on positive portion of Z-axis: **1,0**

Use the SPLINE command to produce a spline running through the points (70,15), (40,20), (10,15), (-30,15), (-60,12), and (-70,10). See Figure 4.6.

    [Draw]   **[Spline]**

    Command: **SPLINE**
    Object/<Enter first point>: **70,15**
    Enter point: **40,20**
    Close/Fit Tolerance/<Enter point>: **10,15**
    Close/Fit Tolerance/<Enter point>: **-30,15**
    Close/Fit Tolerance/<Enter point>: **-60,12**
    Close/Fit Tolerance/<Enter point>: **-70,10**
    Close/Fit Tolerance/<Enter point>: **[Enter]**
    Enter start tangent: **[Enter]**
    Enter end tangent: **[Enter]**

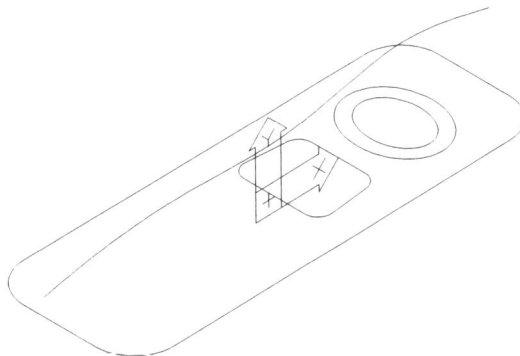

Figure 4.6 Spline created

Set the UCS again using the UCS command. Draw an arc with the ARC command. This arc, together with the spline that you have just created, define the contour of the top surface of the phone. See Figure 4.7.

    [UCS]   **[Z Axis Vector UCS]**

Command: **UCS**
Origin/ZAxis/3point/OBject/View/X/Y/Z/Prev/Restore/Save/Del/?/<World>: **ZA**
Origin point: **0,0,0**
Point on positive portion of Z-axis>: **-1,0**

[Draw]   [Arc 3 Points]

Command: **ARC**
Center/<Start point>: **35,20**
Center/End/<Second point>: **0,30**
End point: **-35,20**

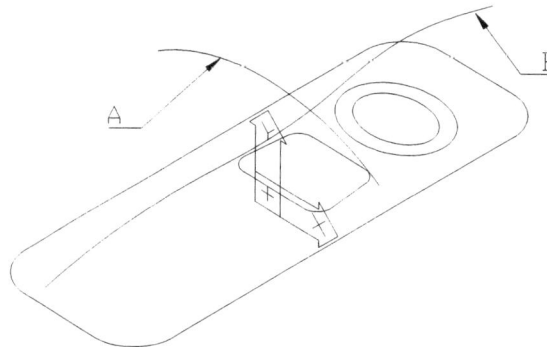

Figure 4.7 Arc created

Run the MOVE command to move the arc. The base point is the quadrant of the arc. The second point is the end of the spline. See Figure 4.8.

[**Modify**]   [**Move**]

Command: **MOVE**
Select objects: **[Select A (Figure 4.7).]**
Select objects: **[Enter]**
Base point or displacement: **QUA** of **[Select A (Figure 4.7).]**
Second point of displacement: **END** of **[Select B (Figure 4.7).]**

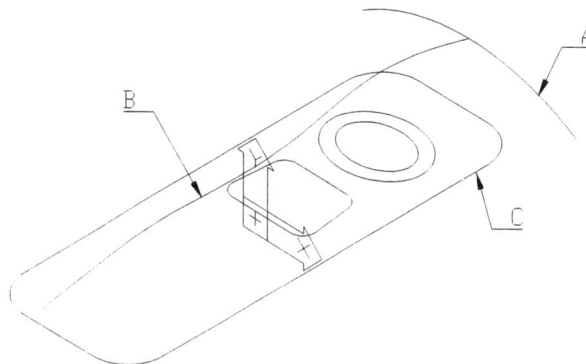

Figure 4.8 Arc moved

Run the GROUP command to put the spline in a group called W4, and the arc in a group called W5.

[Standard Toolbar]   **[Object Group]**

Command: **GROUP**

      [Group name: **W4**
      Selectable: **Yes**
      New:]

Select objects for grouping:
Select objects: **[Select B (Figure 4.8).]**
Select objects: **[Enter]**

      [Group name: **W5**
      Selectable: **Yes**
      New:]

Select objects for grouping:
Select objects: **[Select A (Figure 4.8).]**
Select objects: **[Enter]**

      [OK]

Draw another arc using the ARC command. This arc defines the contour of the side surfaces of the phone. See Figure 4.9.

[Draw]   **[Arc Start Center Angle]**

Command: **ARC**
Center/<Start point>: **END** of **[Select C (Figure 4.8).]**
Center/End/<Second point>: **C**
Center: **@-60,3**
Angle/Length of chord/<End point>: **A**
Included angle: **35**

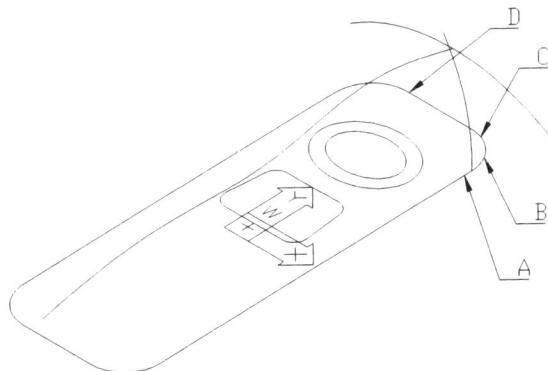

Figure 4.9 Another arc created

Reset the UCS to WORLD using the UCS command. Then, run the ARRAY command and the COPY command to make two more copies of the last arc. See Figure 4.10.

[UCS]   **[World UCS]**

Command: **UCS**
Origin/ZAxis/3point/OBject/View/X/Y/Z/Prev/Restore/Save/Del/?/<World>: **W**

[Copy]   **[Polar Array]**

Command: **ARRAY**
Select objects: **[Select A (Figure 4.9).]**
Select objects: **[Enter]**
Rectangular or Polar array (R/P): **P**
Center point of array: **CEN** of **[Select B (Figure 4.9.)]**
Number of items: **2**
Angle to fill (+=ccw, -=cw) <360>: **90**
Rotate objects as they are copied? <Y> **[Enter]**

[Modify]   **[Copy Object]**

Command: **COPY**
Select objects: **LAST**
Select objects: **[Enter]**
<Base point or displacement>/Multiple: **END** of **[Select C (Figure 4.9).]**
Second point of displacement: **END** of **[Select D (Figure 4.9).]**

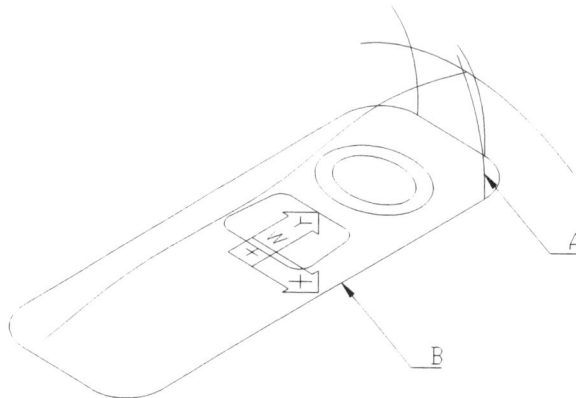

Figure 4.10 Completed wireframes for the mobile phone

You have completed the wireframes for this mobile phone.

4.2   **NURBS Surface Model**

Figure 4.11 shows the completed mobile phone surface model.

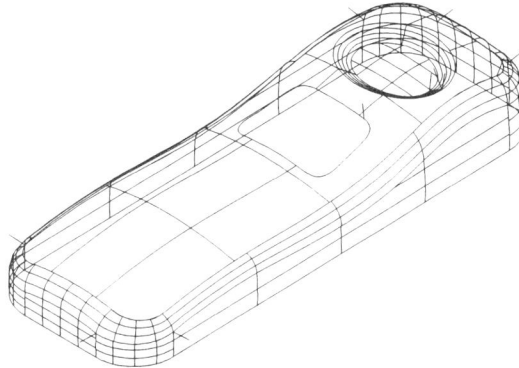

Figure 4.11 Completed surfaces for the mobile phone

The model consists of swept surfaces, fillet surfaces, an offset surface, and a rule surface. First, you will make the swept surfaces. Then, you will form constant fillets, linear variable fillets, and cubical variable fillets between the swept surfaces. Next, you will copy the edges of the fillet surfaces and use the copied edges to trim the top surface.

After completing the main body, you will create an offset surface from the top surface. Using the two ellipses as projection wires, you will project and trim the top surface and the offset surface. Then, you will make a ruled surface to complete the earpiece and use the trapezoidal wire to cut an opening.

Make a layer called SURF with color yellow and set it as the current layer. You will put the surface entities on this layer.

> **<Data>**   **<Layers...>**

Command: **DDLMODES**

Layer	Color
SURF	**yellow**

Current layer: **SURF**

Run the AMSWEEPSF command to create a swept surface. See Figure 4.12.

> **<Surfaces>**   **<Create Surface>**   **<Sweep>**

Command: **AMSWEEPSF**
Select cross sections: **[Select A (Figure 4.10).]**
Select cross sections: **[Enter]**
Select rails: **[Select B (Figure 4.10).]**
Select rails: **[Enter]**

> [Orientation   **Normal**
>   **OK**   ]

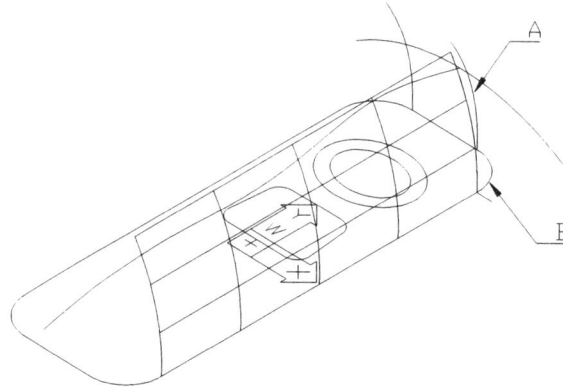

Figure 4.12 Swept surface created

Run the AMSWEEPSF command to create another swept surface. See Figure 4.13.

<Surfaces>   **<Create Surface>**   **<Sweep>**

Command: **AMSWEEPSF**
Select cross sections: **[Select A (Figure 4.12).]**
Select cross sections: **[Enter]**
Select rails: **[Select B (Figure 4.12).]**
Select rails: **[Enter]**

[Orientation   **Normal**
   OK   ]

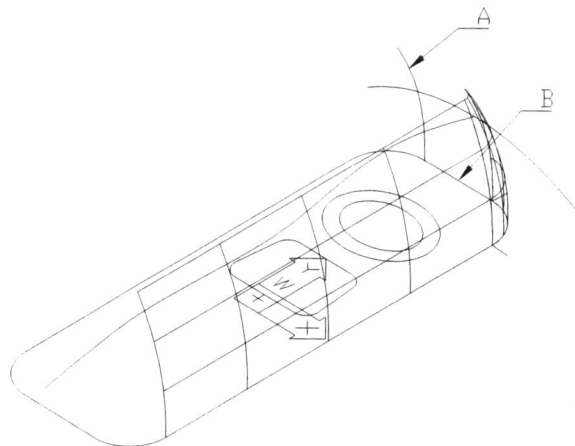

Figure 4.13 Second swept surface created

Run the AMSWEEPSF command again. See Figure 4.14.

<Surfaces>   **<Create Surface>**   **<Sweep>**

Command: **AMSWEEPSF**

Select cross sections: **[Select A (Figure 4.13).]**
Select cross sections: **[Enter]**
Select rails: **[Select B (Figure 4.13).]**
Select rails: **[Enter]**

[Orientation   **Normal**
   OK   ]

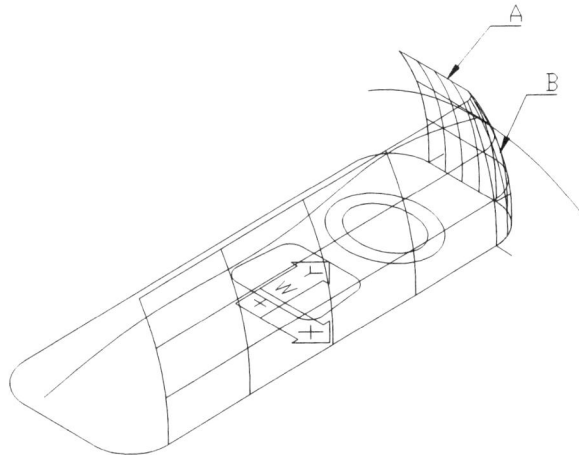

Figure 4.14 Three swept surfaces created

The mobile phone is symmetrical about its center. To create the surfaces of the other side, you will use the MIRROR command to mirror two swept surfaces about the X axis. See Figure 4.15.

[**Copy**]   [**Mirror**]

Command: **MIRROR**
Select objects: **[Select A and B (Figure 4.14).]**
Select objects: **[Enter]**
First point of mirror line: **0,0**
Second point: **1,0**
Delete old objects? <N> **N**

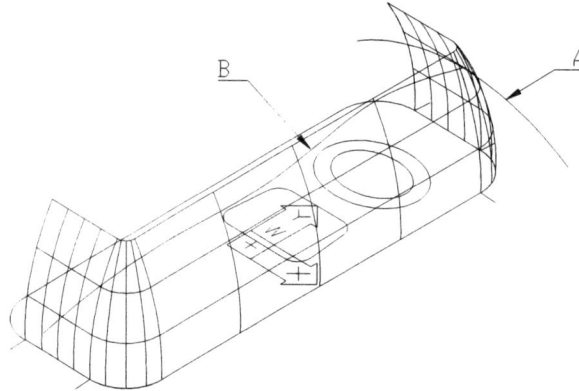

Figure 4.15 Two swept surfaces mirrored

After making three sides and two corners of the phone, run the AMSWEEPSF command on wireframes W3 and W4 to produce a swept surface for the top surface.

\<Surfaces\>   **\<Create Surface\>**   **\<Sweep\>**

Command: **AMSWEEPSF**
Select cross sections: **[Select A (Figure 4.15).]**
Select cross sections: **[Enter]**
Select rails: **[Select B (Figure 4.15).]**
Select rails: **[Enter]**

   [Orientation **Normal**
    OK   ]

To clear the unnecessary entities off the screen, use the LAYER command to turn off the layer WIRE. See Figure 4.16.

   \<Data\>   **\<Layers...\>**

Command: **DDLMODES**

Layer	
WIRE	**Off**

Current layer: **SURF**

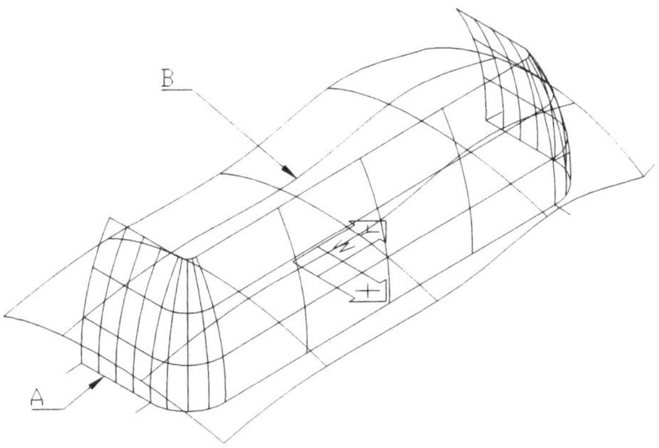

Figure 4.16 Top surface created and layer WIRE turned off

To treat the intersecting surfaces, you will create five fillet surfaces. While filleting, you should trim only the side surfaces and not the top surface. You will use the edges of the fillet surfaces to trim the top surface later. If you trim both surfaces now, you can have unexpected results.

While you are working on this command, you should select the surfaces in sequence according to which surface you want to trim. If you follow the delineation below to select the side surface first and then select the top surface, you should choose to trim the first surface, but not the second.

Run the AMFILLETSF command to create a constant fillet. See Figure 4.17.

\<Surfaces\>   \<Create Surface\>   \<Fillet...\>

Command: **AMFILLETSF**
Select first surface: **[Select A (Figure 4.16).]**
Select second surface: **[Select B (Figure 4.16).]**

[Trim:
First surface: **Yes**
Second surface: **No**
Fillet Radius: **5**
OK   ]

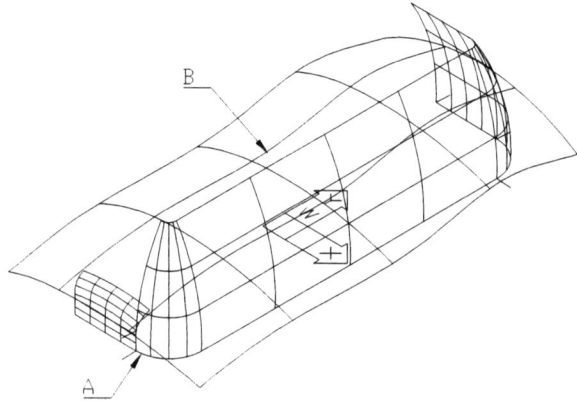

Figure 4.17 Constant fillet created

The second fillet surface is a variable fillet. The fillet radius varies linearly from 5 units to 4 units. In selecting the surfaces, you should take great care to ensure that you select a point near the edge of the surface where you put the start fillet radius. See Figure 4.18.

<Surfaces>   <Create Surface>   <Fillet...>

Command: **AMFILLETSF**
Select first surface: **[Select A (Figure 4.17).]**
Select second surface: **[Select B (Figure 4.17).]**

	[Trim:	
	First surface:	**Yes**
	Second surface:	**No**
	Fillet Type:	
	Variable:	**Linear**
	First edge:	**5**
	Second edge:	**4**
	OK	]

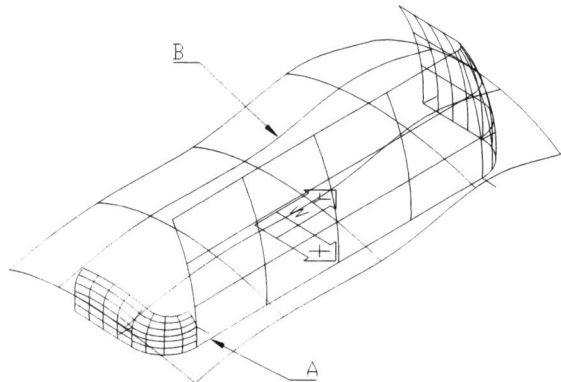

Figure 4.18 Linear variable fillet created

Create the third fillet. It is a cubical variable fillet that changes from 4 units to 6 units. See Figure 4.19.

<Surfaces>   **<Create Surface>**   **<Fillet...>**

Command: **AMFILLETSF**
Select first surface: **[Select A (Figure 4.18).]**
Select second surface: **[Select B (Figure 4.18).]**

[Trim:
  First surface: **Yes**
  Second surface: **No**
  Fillet Type:
  Variable: **Cubic**
  First edge: **4**
  Second edge: **6**
  OK   ]

Before you fillet the remaining edges, use the VPOINT command to switch to a new viewing position.

<View>   **<3D Viewpoint>**   **<Vector>**

Command: **VPOINT**
Rotate/<View point>: **R**
Enter angle in XY plane from X axis: **45**
Enter angle from XY plane: **35**

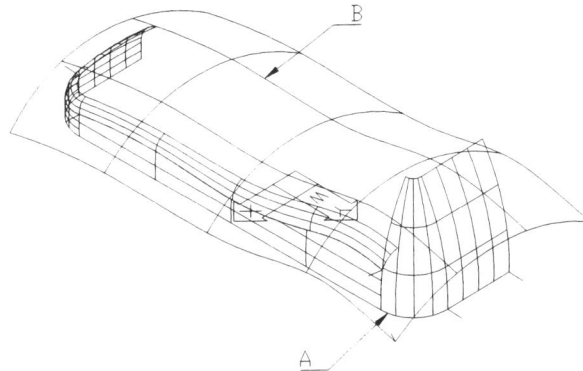

Figure 4.19 Cubical variable fillet created and viewed from another angle

Run the AMFILLETSF command to create a cubical variable fillet. See Figure 4.20.

<Surfaces>   **<Create Surface>**   **<Fillet...>**

Command: **AMFILLETSF**
Select first surface: **[Select A (Figure 4.19).]**
Select second surface: **[Select B (Figure 4.19).]**

[Trim:
  First surface: **Yes**
  Second surface: **No**
  Fillet Type:
  Variable: **Cubic**
  First edge: **6**
  Second edge: **7**
  OK   ]

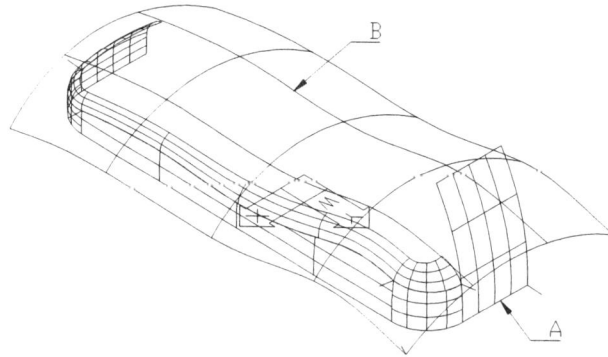

Figure 4.20 Another cubical variable fillet created

Run the AMFILLETSF command to make a constant radius fillet. See Figure 4.21.

<Surfaces>   **<Create Surface>**   **<Fillet...>**

Command: **AMFILLETSF**
Select first surface: **[Select A (Figure 4.20).]**
Select second surface: **[Select B (Figure 4.20).]**

[Trim:
  First surface: **Yes**
  Second surface: **No**
  Fillet Radius: **7**
  OK   ]

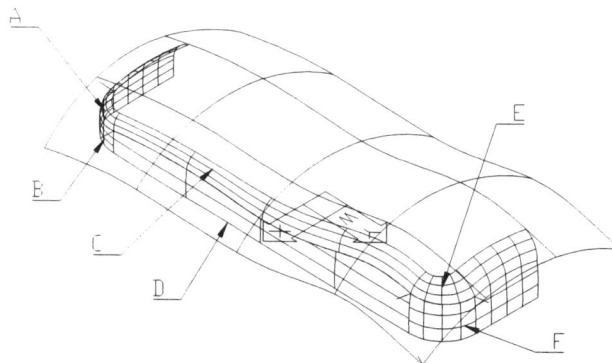

Figure 4.21 Constant fillet created

Because the phone is symmetrical about its Y axis, you will run the MIRROR command to mirror three swept surfaces and three fillet surfaces to complete the sides and fillets. See Figure 4.22.

[Copy]   **[Mirror]**

Command: **MIRROR**
Select objects: **[Select A, B, C, D, E, and F (Figure 4.21).]**
Select objects: **[Enter]**
First point of mirror line: **0,0**
Second point: **0,1**
Delete old objects? <N> **N**

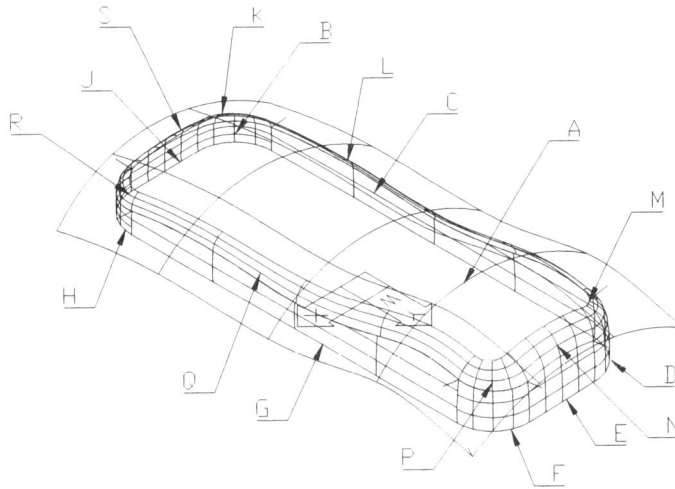

Figure 4.22 Swept and fillet surfaces mirrored

The side surfaces and the fillet surfaces are completed. Use the GROUP command to put the top surface in an entity group S1, the eight side swept surfaces in an entity group S2, and the eight fillet surfaces in an entity group S3.

[Standard Toolbar]   **[Object Group]**

Command: **GROUP**

       [Group name: **S1**
       Selectable: **Yes**
       New:]

Select objects for grouping:
Select objects: **[Select A (Figure 4.22).]**
Select objects: **[Enter]**

       [Group name: **S2**
       Selectable: **Yes**
       New:]

Select objects for grouping:
Select objects: **[Select B, C, D, E, F, G, H, and J (Figure 4.22).]**
Select objects: **[Enter]**

  [Group name: **S3**
   Selectable: **Yes**
   New:]

Select objects for grouping:
Select objects: **[Select K, L, M, N, P, Q, R, and S (Figure 4.22).]**
Select objects: **[Enter]**

  [OK]

To trim the top surface, you need to copy the edges from the fillet surfaces. To copy the correct edges, it is better to clear the unrelated objects from the screen. Run the AMVISIBLE command to hide all the entities except the entity group S3, the fillet surfaces. See Figure 4.23.

  <Surfaces> **<Object Visibility...>**

Command: **AMVISIBLE**

   [Hide
   All
   Except   ]

Select objects to remain visible: **GROUP**
Enter group name: **S3**
Select objects to remain visible: **[Enter]**

  [OK]

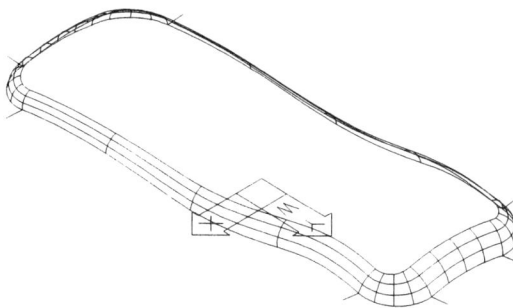

Figure 4.23 All entities hidden except S3

Before copying the edges, set the display to show the top view by using the accelerator key [5].

Command: **5**

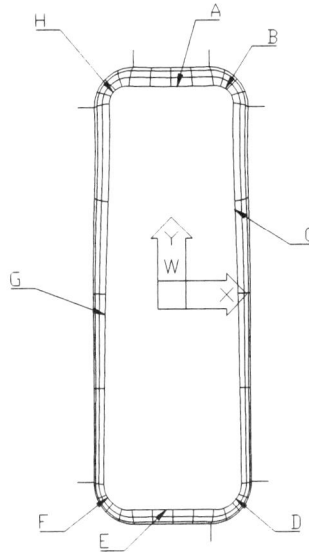

Figure 4.24 Display set to top view

To put the copied edges on a separate layer, make a layer called UTY with color blue, and set it as the current layer.

> **<Data>**   **<Layers...>**

Command: **DDLMODES**

Layer	Color
UTY	**blue**

Current layer: **UTY**

Use the AMEDGE command to copy the upper edges of the fillet surfaces. As you copy, you will find some blue wires appear at the edges.

> **<Surfaces>**   **<Create Wireframe>**   **<Copy Edge>**

Command: **AMEDGE**
(Output = Polyline)
Copy edge/Output/Show nodes/Untrim/<Extract loop>: **COPY**
Select surface edge: **[Select A, B, C, D, E, F, G, and H (Figure 4.24).]**
Select surface edge: **[Enter]**

In order to see the copied edges clearly, run the AMVISIBLE command to hide the fillet surfaces. To use the copied edges to trim the top surface, you should also unhide the entity group S1. See Figure 4.25.

> **<Surfaces>**   **<Object Visibility...>**

Command: **AMVISIBLE**

[Hide
   Select   ]

Select objects to hide: **GROUP**
Enter group name: **S3**
Select objects to hide: **[Enter]**

[Unhide
   Select   ]

Select objects to unhide: **GROUP**
Enter group name: **S1**
Select objects to unhide: **[Enter]**

[OK]

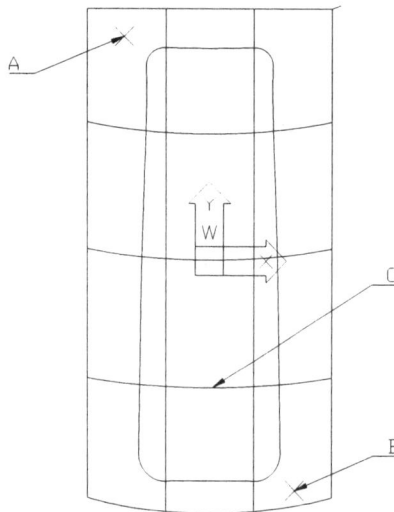

Figure 4.25 Edges copied, S3 hidden, and S1 unhidden

Using the copied edges as projection wires, run the AMPROJECT command to trim the top surface S1. See Figure 4.26.

<Surfaces>   **<Edit Surface>**   **<Project Trim...>**

Command: **AMPROJECT**
Select wires to project: **[Select A (Figure 4.25).]**
Other corner: **[Select B (Figure 4.25).]**
Select wires to project: **[Enter]**
Select target surfaces: **[Select C (Figure 4.25).]**
Select target surfaces: **[Enter]**

    [Direction: **Vector prompts**
     Output type: **Trim surface**
     OK   ]

Viewdir/Wire/X/Y/Z/<Start point>: **Z**

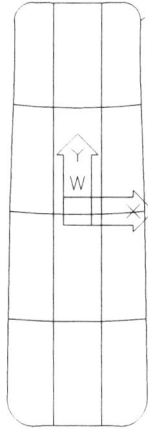

Figure 4.26 Top surface S1 trimmed

After trimming, the main body of the mobile phone surface model is complete. To continue, you will create two more features on it: the earpiece and the display panel opening.

Switch to an isometric view using the accelerator key [8].

Command: **8**

Set the current layer to SURF.

<Data>   <Layers...>

Command: **DDLMODES**

Current layer: **SURF**

To make the earpiece, you have to derive an offset surface. The new surface should reside on the negative Z direction of the top surface. Before you use the AMOFFSETSF command, examine the normal direction of the top surface. It should point towards the positive Z direction. If not, use the AMEDITSF command to reverse its normal direction.

<Surfaces>   <Edit Surface>   <Flip Normal>

To reiterate, the normal should point upwards.
To keep the original top surface, set the system variable DELOBJ to zero.

Command: **DELOBJ**
New value for DELOBJ: **0**

Now use the AMOFFSETSF command to make an offset copy of the top surface. See Figure 4.27.

<Surfaces>   <Create Surface>   <Offset>

Command: **AMOFFSETSF**
Select surfaces: **GROUP**
Enter group name: **S1**
Select surfaces: **[Enter]**
Offset distance: **-3**

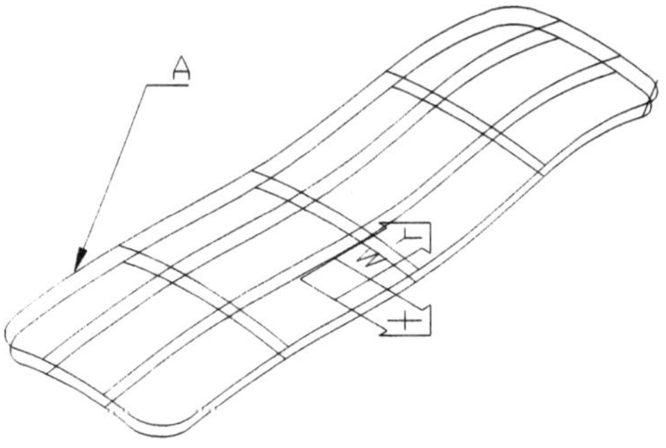

Figure 4.27 Offset surface created

After offsetting, you should have two surfaces on the screen: the original top surface and the offset surface in the negative Z direction.

Turn on the layer WIRE.

<Data>   **<Layers...>**

Command: **DDLMODES**

Layer	
WIRE	**ON**

Current layer: **SURF**

To create the earpiece of the mobile phone, you will use the larger ellipse to trim the top surface and the smaller ellipse to trim the offset surface.

When you use the AMPROJECT command to trim, select the place you want to retain. For the top surface, you should select near the edge. Refer to Figure 4.27.

<Surfaces>   **<Edit Surface>**   **<Project Trim...>**

Command: **AMPROJECT**
Select wires to project: **GROUP**
Enter group name: **W1**
Select wires to project: **[Enter]**
Select target surfaces: **[Select A (Figure 4.27).]**
Select target surfaces: **[Enter]**

> [Direction: **Vector prompts**
> Output type: **Trim surface**
> **OK**   ]

Viewdir/Wire/X/Y/Z/<Start point>: **Z**

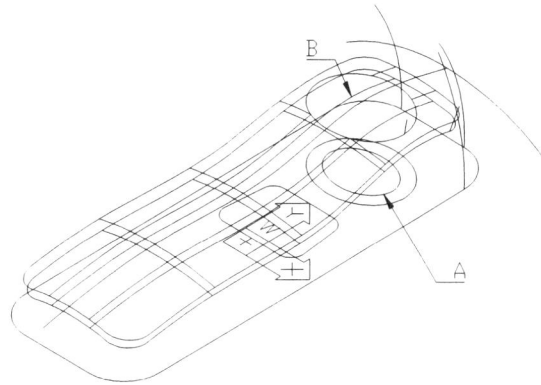

Figure 4.28 Top surface projected and trimmed

Run the AMPROJECT command on the offset surface. You should select within the projected zone of the smaller ellipse. Refer to Figure 4.28.

<**Surfaces**>   <**Edit Surface**>   <**Project Trim...**>

Command: **AMPROJECT**
Select wires to project: **GROUP**
Enter group name: **W2**
Select wires to project: **[Enter]**
Select target surfaces: **[Select A (Figure 4.28).]**
Select target surfaces: **[Enter]**

        [Direction: **Vector prompts**
         Output type: **Trim surface**
         OK   ]

Viewdir/Wire/X/Y/Z/<Start point>: **Z**

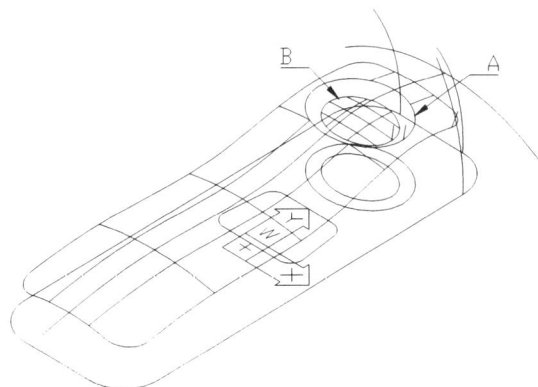

Figure 4.29 Offset surface projected and trimmed

To complete the earpiece, use the AMRULE command to produce a ruled surface that runs between the trimmed edges of the top surface and the offset surface. See Figure 4.30.

<Surfaces>   <Create Surface>   <Rule>

Command: **AMRULE**
Select first wire: **[Select A (Figure 4.29.]**
Select second wire: **[Select B (Figure 4.29).]**

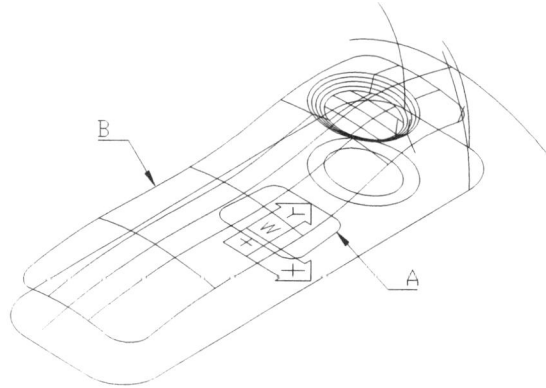

Figure 4.30 Ruled surface between the top and offset surfaces created

The earpiece of the mobile phone is complete.

To cut an opening for the display panel, use the AMPROJECT command to trim the top surface again. See Figure 4.31.

<Surfaces>   <Edit Surface>   <Project Trim...>

Command: **AMPROJECT**
Select wires to project: **[Select A (Figure 4.30).]**
Select wires to project: **[Enter]**
Select target surfaces: **[Select B (Figure 4.30).]**
Select target surfaces: **[Enter]**

　　　　　[Direction: **Vector prompts**
　　　　　 Output type: **Trim surface**
　　　　　 OK   ]

Viewdir/Wire/X/Y/Z/<Start point>: **Z**

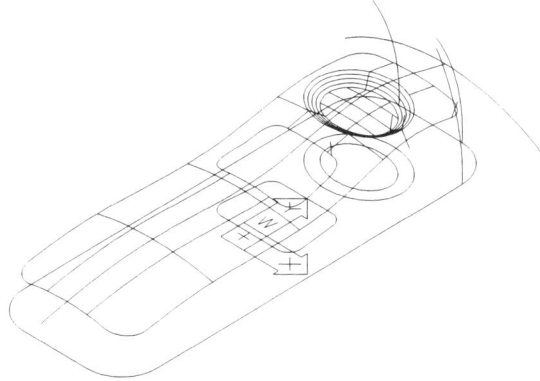

Figure 4.31 Top surface trimmed again

Turn off the layer WIRE, and unhide all the entities. See Figure 4.11 again.

<Data>   **<Layers...>**

Command: **DDLMODES**

WIRE	Off

<Surfaces>   **<Object Visibility...>**

Command: **AMVISIBLE**

    [Unhide
     All
     OK   **]**

As an additional exercise, try to cut the 15 button openings on this model by yourself. Figure 4.32 shows the suggested dimensions for the buttons, and Figure 4.1 shows the completed model with the button openings.

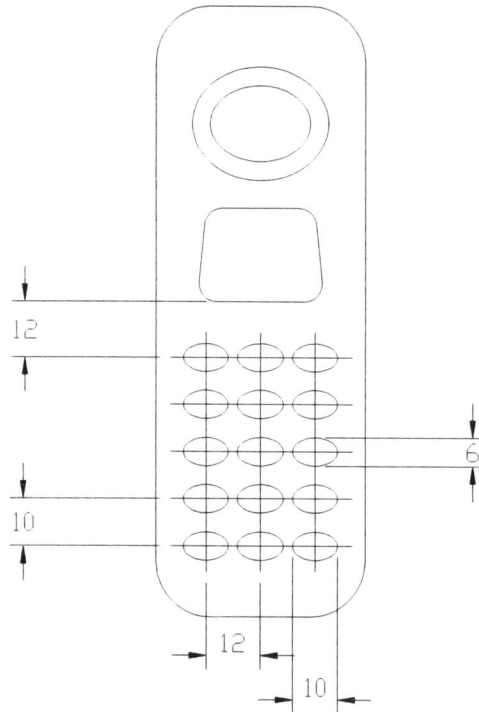

Figure 4.32 Suggested button dimensions

Save your work to a file with the SAVE or SAVEAS command.

Command: **SAVE**

4.3   Utilities

NURBS surface models are ideal 3D free-form objects in a computer for use in manufacturing processes. Based on a NURBS surface model, you can output flow lines, parting lines, cross sections, and augmented lines.

Set the current layer to UTY. You will create flow lines, parting lines, cross sections, and augmented lines on this layer.

<Data>   <Layers...>

Command: **DDLMODES**

Current layer: **UTY**

Flow lines are U- and V-wire mesh of a surface in two orthogonal directions. In making the mesh, you can specify any number of U- and V-wires from a surface, regardless of the current UV display line number or UV patch line number.

Run the AMFLOW command to see how the flow lines look. See Figure 4.34.

\<Surfaces\>   **\<Create Wireframe\>**   **\<Flow...\>**

Command: **AMFLOW**
Select surfaces: **GROUP**
Enter group name: **S1**
Select surfaces: **[Enter]**

       [Flow Type
       UV Wires
       U Wires   **10**
       V Wires   **4**
       Save   **Yes**
       Preview   ]

Enter RETURN to continue: **[Enter]**

    [OK]

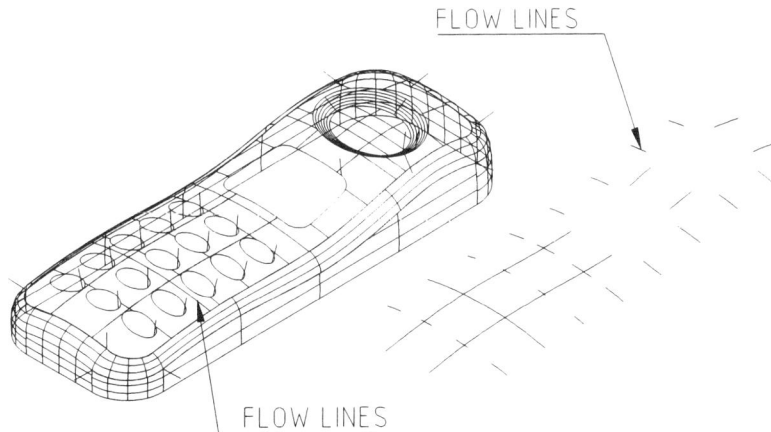

Figure 4.33 Flow lines created and copied aside

As you can see, the U- and V-wire mesh does not include the edges of the surface. If you need to have the edges, you should use the AMEDGE command.

Like the edges of a surface, you can treat the resulting flow lines as ordinary wires for defining a surface, trimming a surface, or projecting another wire on a surface. If you need the wires for these or other purposes, you should save them. In addition to setting the wire mesh density, you can output offset flow lines. Offset flow lines are much the same as flow lines generated from an offset surface. If you want a set of wires offset at a distance from a given surface, you do not have to make an offset surface for outputting flow lines. Instead, you can specify an offset distance while making the flow lines.

When you use a NURBS surface model to make molds, you need a parting line to split the surface model into two halves. To obtain a parting line, you simply use the AMPARTLINE command. See Figure 4.34.

\<Surfaces\>   **\<Create Wireframe\>**   **\<Parting Line\>**

Command: **AMPARTLINE**
Select surfaces: **GROUP**
Enter group name: **S2**
Select surfaces: **[Enter]**
Direction: Viewdir/Wire/X/Y/Z/<Start point>: **Z**

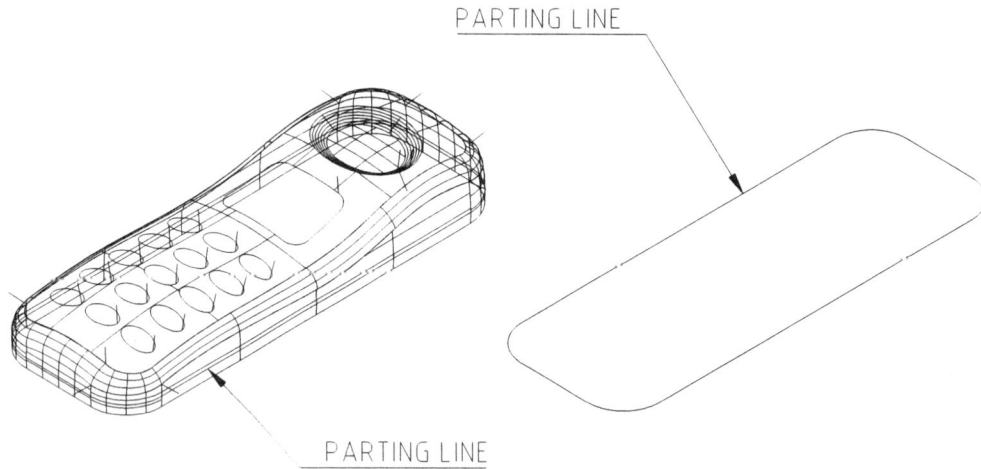

Figure 4.34 Parting line created and copied aside

Flow lines are wires on a single surface. If you want to generate a cross section across a set of surfaces of a surface model, then you should use the AMSECTION command. This command will output a series of wires at regular intervals. In addition to using the sections as wires, you can use them to verify and check your design, and as tool paths for machining the surface.

Set the UCS to a new orientation. Then, run the AMSECTION command to generate a section line across the surface model. See Figure 4.35.

    [UCS]   **[Z Axis Vector UCS]**

Command: **UCS**
Origin/ZAxis/3point/OBject/View/X/Y/Z/Prev/Restore/Save/Del/?/<World>: **ZA**
Origin point: **0,0,0**
Point on positive portion of Z-axis: **1,0**

    <Surfaces>   **<Create Wireframe>**   **<Section Cuts...>**

Command: **AMSECTION**
Select surfaces: **ALL**
Select surfaces: **[Enter]**

        [Section Type   **Single**
        Initial Plane   **UCS Plane**
        Save   **Yes**
        OK   **]**

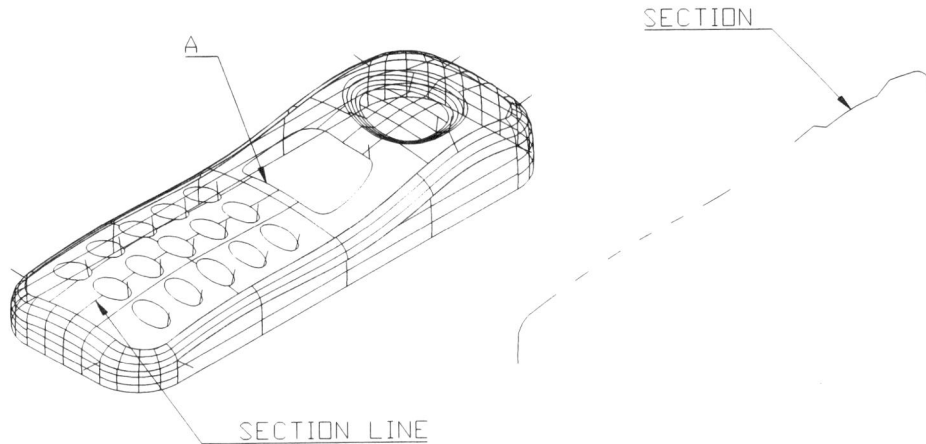

Figure 4.35 Section lines created and copied

Regarding tool paths, advanced five-axis machines have two more cutting tool motions available in addition to X, Y, and Z movements. These two motions are rotational motions about the X axis and the Y axis. Naturally, simple 3D tool paths from the U and V flow lines or the section lines are inadequate to these machines.

To meet such need, you will generate augmented lines from the surface model. Augmented lines are polylines with normal vectors along the polylines. Create augmented lines by using the AMAUGMENT command. See Figure 4.36.

<Surfaces>   <Create Wireframe>   <Augmented Lines>

```
Command: AMAUGMENT
(Angle = 0, Distance = 0, Spacing = Optimal)
Angle/Distance/Spacing/<sElect surface wire>: S
Step/Vertices/Optimal: S
Step: 2
(Angle = 0, Distance = 0, Spacing = Step)
Angle/Distance/Spacing/<sElect surface wire>: [Select A (Figure 4.35).]
Angle/Distance/Spacing/<sElect surface wire>: [Enter]
```

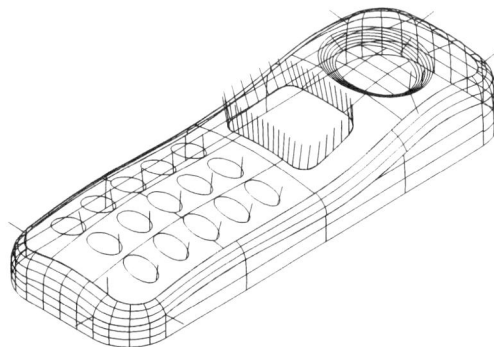

Figure 4.36 Augmented lines created

If you compare your drawing with Figure 4.36, the length of the augmented lines might not be the same. To change their length, you can use the AMSURFVARS command before making the augmented lines, or you can edit the augmented lines later.

The default directions of the vectors on the augmented lines is 0°, i.e. normal to the surface. To meet practical needs, you may edit these vectors.

Run the AMEDITAUG command to change the vector length, and to rotate the vectors. See Figure 4.37.

<Surfaces>   <Edit Wireframe> <Vector Length>

Command: **AMEDITAUG**
Add vectors/Blend/Copy/Normal length/Rotate/Twist/<eXit>: **N**
Select augmented lines: **[Select the augmented lines (Figure 4.36).]**
Select augmented lines: **[Enter]**
Normal length: **10**

<Surfaces>   <Edit Wireframe> <Rotate Vectors>

Command: **AMEDITAUG**
Add vectors/Blend/Copy/Normal length/Rotate/Twist/<eXit>: **R**
Plane/<Angle>: **A**
Angle <90>: **45**
All/Range/<Select vector>: **ALL**
Select augmented line: **[Select the augmented lines (Figure 4.36).]**
Select augmented line: **[Enter]**

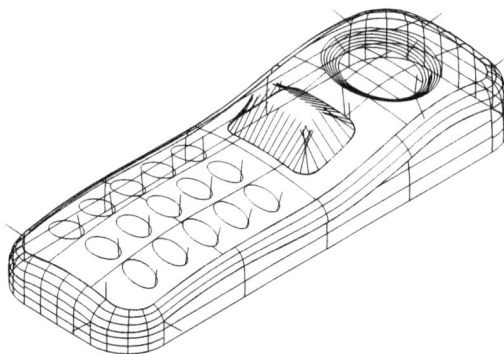

Figure 4.37 Augmented lines edited

A surface itself does not have a thickness. Therefore, you cannot ask the computer to compute the mass property of a surface unless you have assigned a thickness to it.

Suppose the thickness of this mobile phone casing is 2 units and the density of the material used to make it is 1.5. Run the AMSURFPROP command to find out the properties of this model shell.

<Surfaces>   <Utilities>   <Mass Properties>

```
Command: AMSURFPROP
Select surfaces: [Pick all the surfaces.]
Select surfaces: [Enter]
(Density = 1, Thickness = 0, tYpe = Shell)
Density/Thickness/tYpe/<Calculate properties>: T
Thickness: 2
(Density = 1, Thickness = 2, tYpe = Shell)
Density/Thickness/tYpe/<Calculate properties>: D
Density: 1.5
(Density = 1.5, Thickness = 2, tYpe = Shell)
Density/Thickness/tYpe/<Calculate properties>: Y
Shell/<Enclosed model>: S
Density/Thickness/tYpe/<Calculate properties>: [Enter]
```

Depending on the setting of the system variable CMDDIA, AutoSurf lists the result at the prompt line or in the dialog box. The following messages appear at the prompt line if the CMDDIA setting is 0.

```
---------------- SURFACES ----------------

Type:          Shell
Thickness:     2
Density:       1.5
Area:          9160.2432
Mass:          27480.7296
```

If the CMDDIA setting is 1, a dialog box similar to the one shown in Figure 4.38 appears.

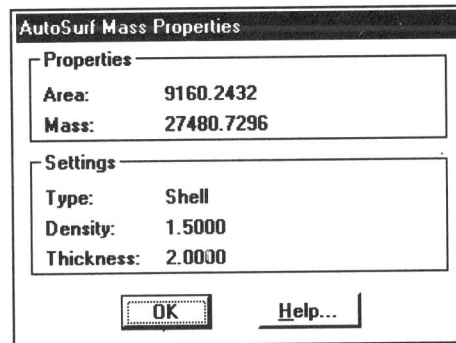

Figure 4.38 AutoSurf Mass Properties dialog box

In pursuing a project that involves making a surface model, you might have to input data from or output data to other applications. Depending on the compatibility of other applications with AutoSurf, you can use the AMMODIN command or the IGESIN command for inputting, and use the AMMODOUT command or the IGESOUT command for outputting.

The AMMODIN command and the AMMODOUT command can be selected from the Surface pull-down menu.

<Surface>   **<Utilities>**   **<.MOD In...>**

Command: **AMMODIN**

\<Surface\>   **\<Utilities\>**   **\<.MOD Out...\>**

Command: **AMMODOUT**

The IGESIN command and the IGESOUT command can be selected from the Mtools pull-down menu. See Figure 4.39 and Figure 4.40.

\< Mtools \>   **\<IGES In...\>**

Command: **IGESIN**

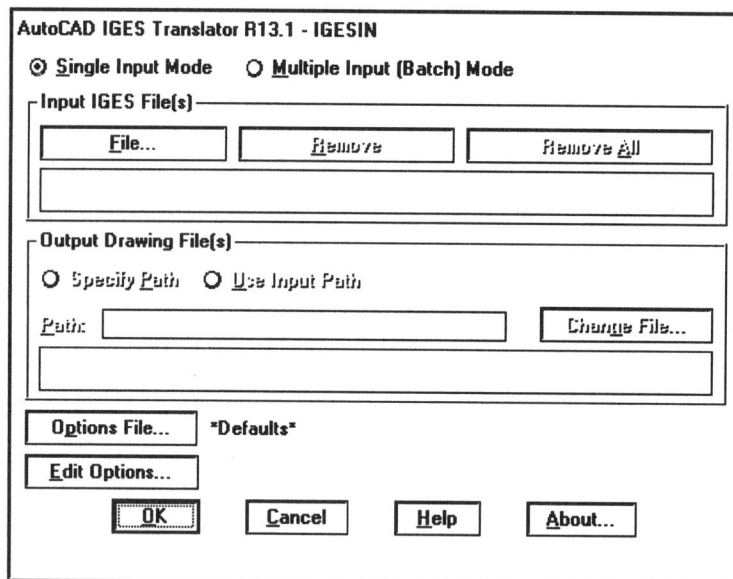

Figure 4.39 IGES file import

\< Mtools \>   **\<IGES Out...\>**

Command: **IGESOUT**

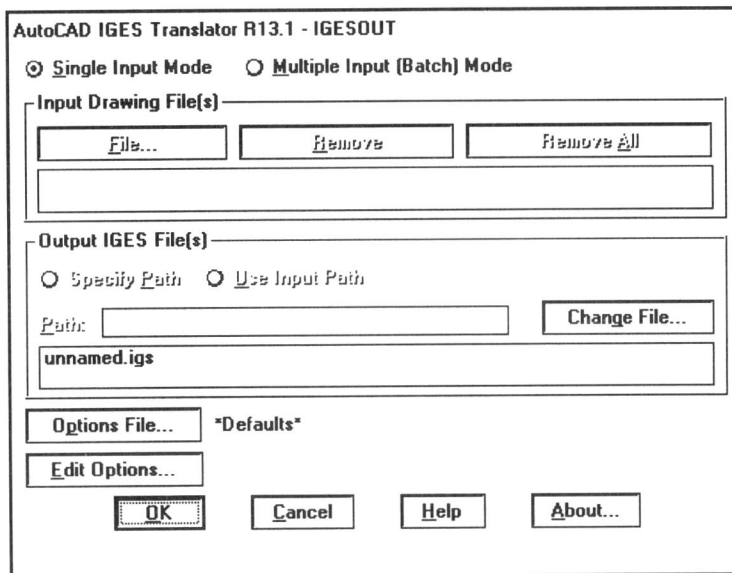

Figure 4.40 IGES file output

4.4 Summary

In this chapter, you have practiced the following AutoSurf commands, in addition to those commands you learned in the previous chapters:

AMAUGMENT	AMEDITAUG	AMFLOW
AMMODIN	AMMODOUT	AMPARTLINE
AMSECTION	AMSURFPROP	IGESIN
IGESOUT		

For a brief explanation of these commands, refer to the appendix of this book.

In carrying out this project, you have practiced how to create wireframes and surfaces of a model, and you have learned how to use the utility commands. By now, you should be able to create flow lines, section lines, parting lines, and augmented lines from a surface model. In addition, you should be able to use the output and input commands.

In the next chapter, you will use a different approach to make a surface model. Instead of creating each surface separately, you will construct a native solid model, use a surface to cut the surface, and convert the native solid to a set of surfaces.

4.5 Exercise

To further enhance your knowledge of surface modeling and surface utilities, you will create the surface model of a joy pad. Figure 4.41 shows a four-viewport display of the completed model.

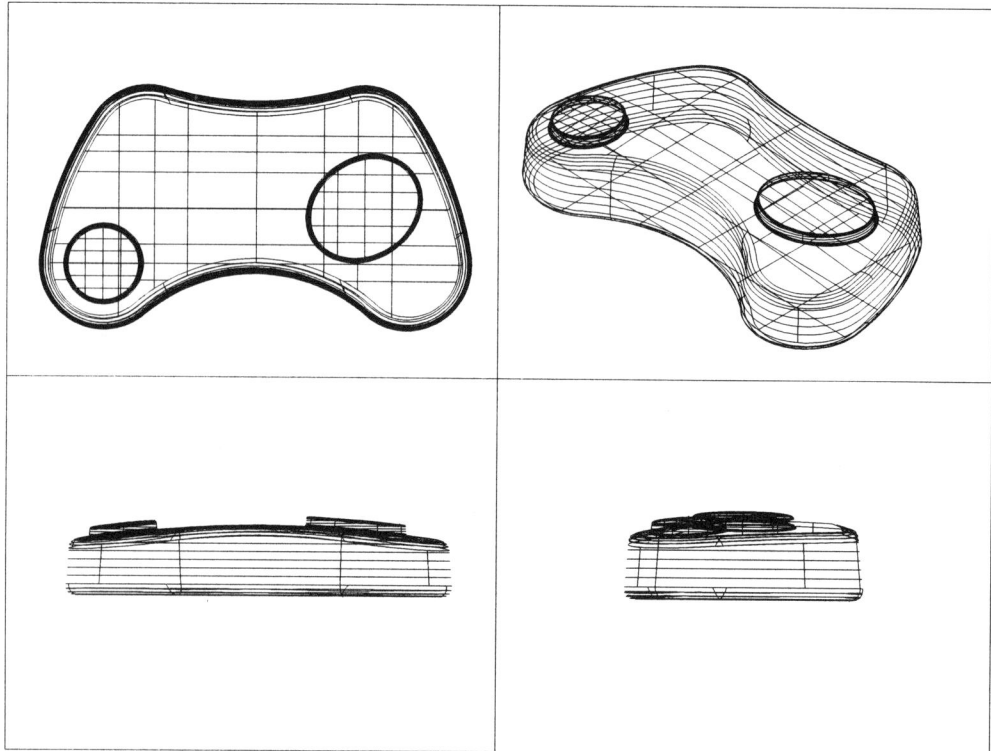

Figure 4.41 Completed surface model of the joy pad

Start a new drawing. Make two additional layers, WIRE and SURF. Set layer WIRE as the current layer.

Before starting to create entities on your drawing, place the UCS icon at the origin position and set AutoSurf preferences by specifying a model size of 200 millimeters.

As shown in Figure 4.42, create eight arc segments. The center position of the lower-left R30 arc is (50,50).

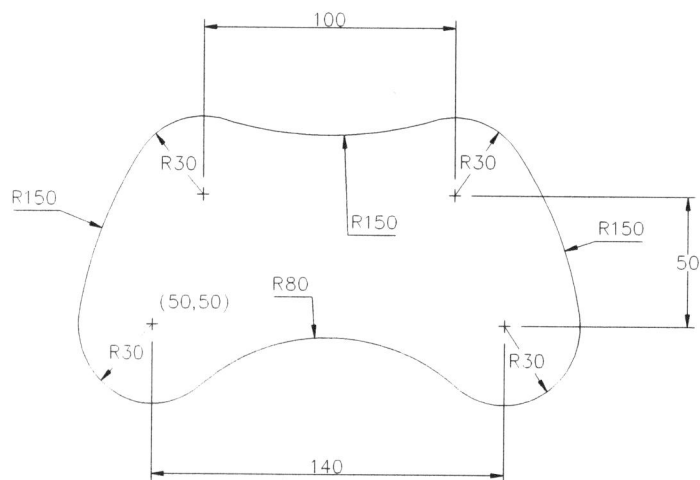

Figure 4.42 Eight arcs created

Join the eight arcs to form a spline. Then, create a circle and an ellipse. See Figure 4.43.

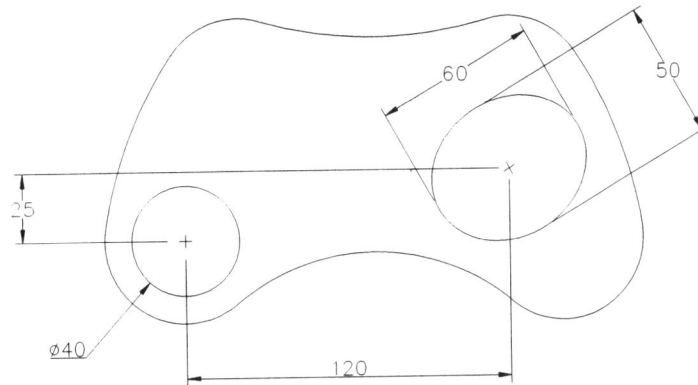

Figure 4.43 Circle and ellipse created

Create three splines, as shown in Figure 4.44. The wireframes are complete.

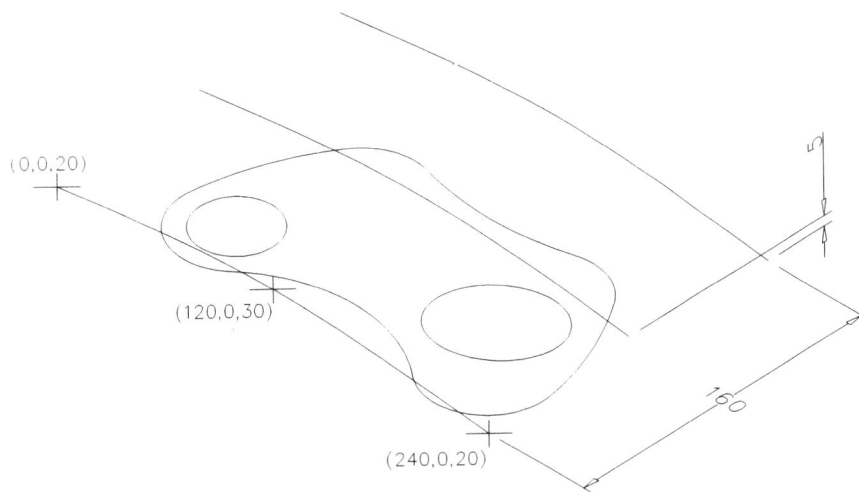

Figure 4.44 Three splines created

Set the current layer to SURF. Then, create a trimmed planar surface and an extruded surface as shown in Figure 4.45.

The distance of extrusion is 50 units. The draft angle is -5°.

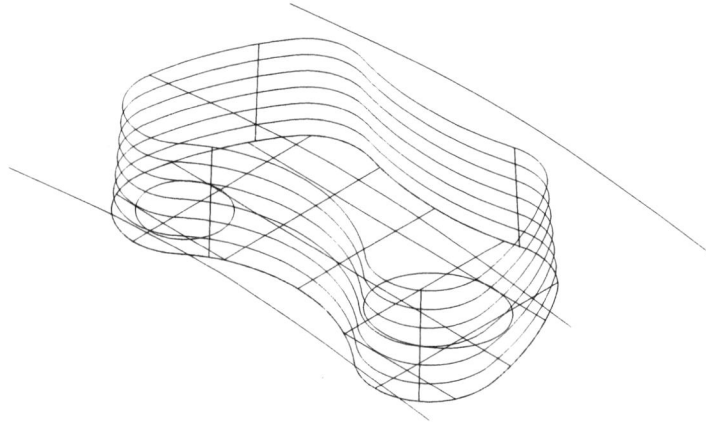

Figure 4.45 Trimmed planar surface and an extruded surface created

Form a fillet surface between the extruded surface and the trimmed planar surface. The fillet radius is 5 units. See Figure 4.46.

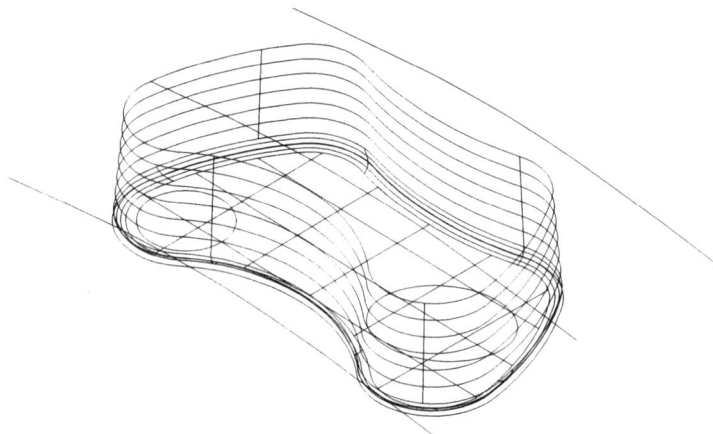

Figure 4.46 Fillet surface created between the extruded surface and the trimmed planar surface

Using the three splines as U-lines, create a lofted U surface. While making the surface, align the directions, choose a smooth curve fit, and respace the curve ends.

After making the lofted U surface, turn off the layer WIRE. See Figure 4.47.

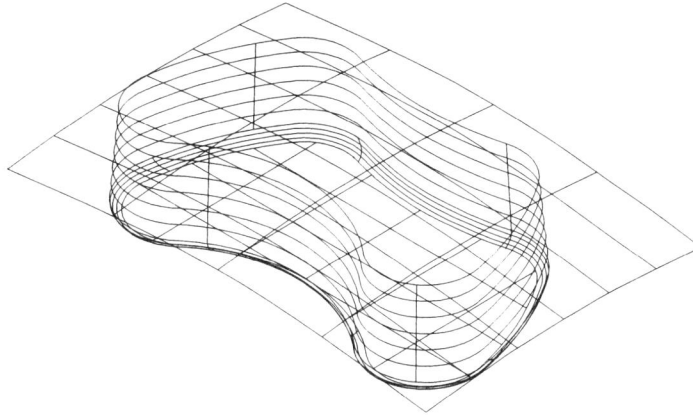

Figure 4.47 Lofted U surface created and layer WIRE turned off

Create a fillet surface of 5 units radius between the extruded surface and the lofted U surface. See Figure 4.48.

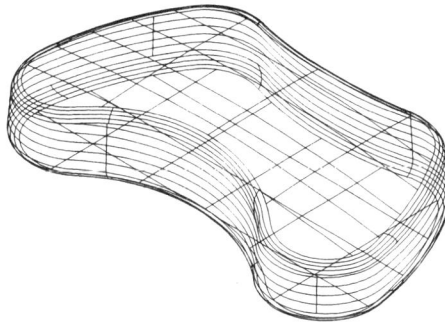

Figure 4.48 Fillet surface formed between the extruded surface and the lofted U surface

Check the normal direction of the lofted U surface. If it is not pointing upwards, flip its direction. Then, set the system variable DELOBJ to 0, and make an offset surface of the lofted U surface at a distance of 5 units upward. See Figure 4.49.

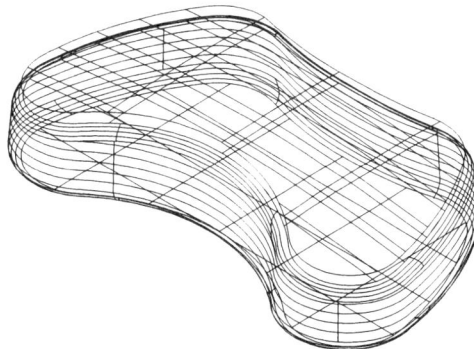

Figure 4.49 Offset surface created

Hide all the surfaces except the lofted U surface and the offset surface. Turn on the layer WIRE. See Figure 4.50.

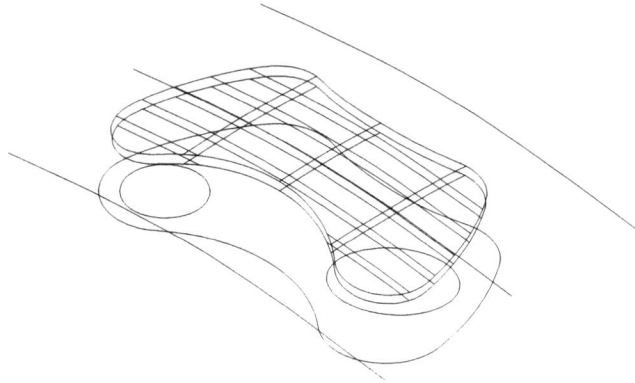

Figure 4.50 Some surfaces hidden and layer WIRE turned on

Make a second copy of the offset surface at the same location. Then, hide the lofted U surface.

Switch to a two-viewport screen display. Then, extrude the circle and the ellipse for a distance of 90 units and -5° draft angle. Turn off the layer WIRE. See Figure 4.51.

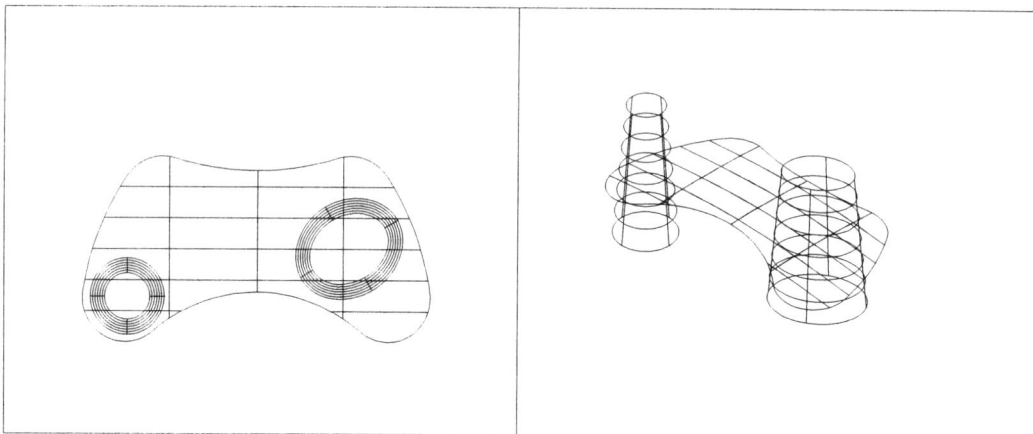

Figure 4.51 Second offset surface created, lofted U surface hidden, and two objects extruded

You should have two identical offset surfaces at the same location. Form a fillet surface of 1 unit between one offset surface and the extruded circle. Then, form another fillet surface of 1 unit between the second offset surface and the extruded ellipse. See Figure 4.52.

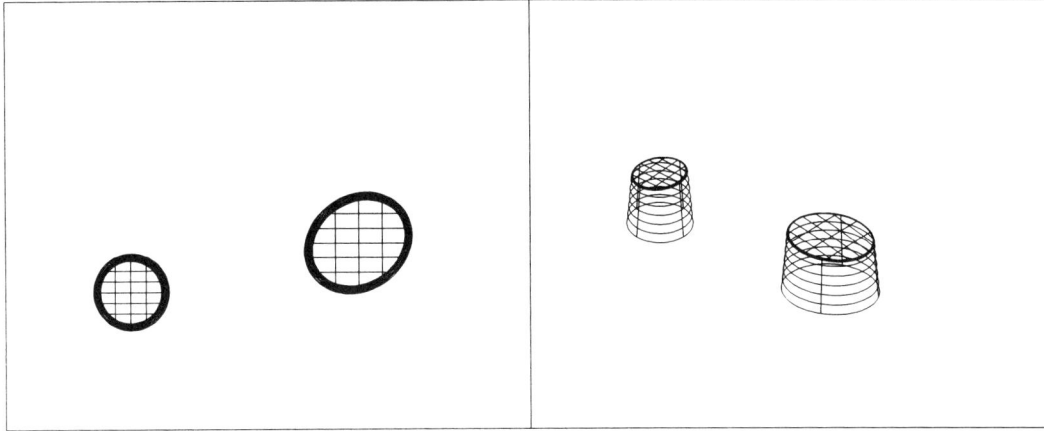

Figure 4.52 Two fillet surfaces formed

Switch to a single viewport display, and set an isometric view. Then, unhide the lofted U surface. See Figure 4.53.

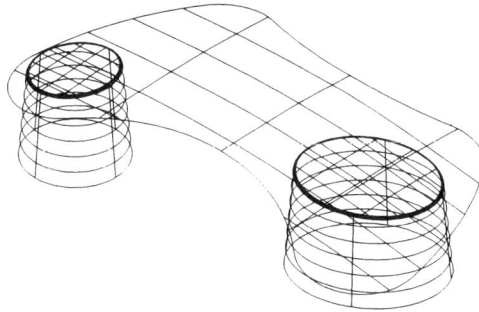

Figure 4.53 Lofted U surface unhidden

As shown in Figure 4.54, form two fillet surfaces of 1 unit radius — one between the extruded circle and the lofted U surface, and the other between the extruded ellipse and the lofted U surface.

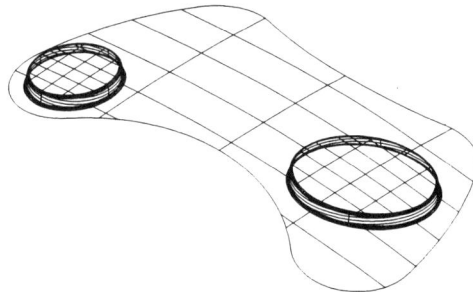

Figure 4.54 Two fillet surfaces created

Unhide all the surfaces. The surface model of the joy pad is complete.

Making use of the completed surface model, output a series of section lines at a 30 units interval. See Figure 4.55.

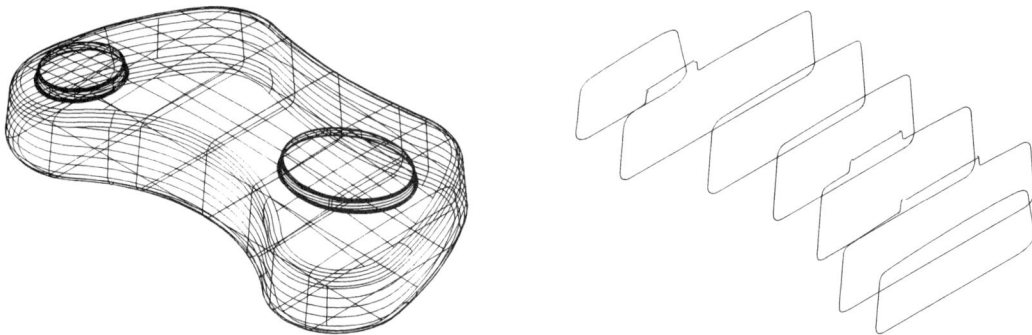

Figure 4.55 Section lines created and copied

Output an augmented line of 10 units along the edge of the lofted U surface. The spacing in steps is 5 units. See Figure 4.56.

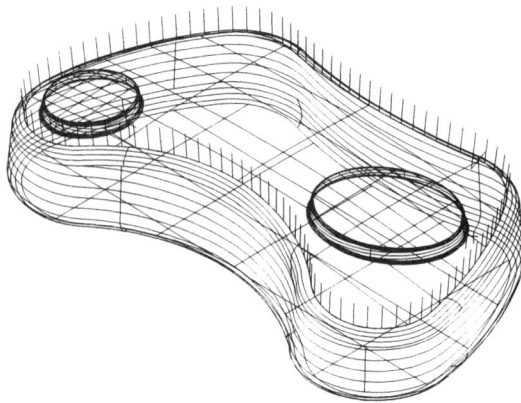

Figure 4.56 Augmented lines created

Save your drawing.

Chapter 5

Inter-Operation and Assembly

Being a component of the Mechanical Desktop Package, AutoSurf Release 3 allows the inter-operation of NURBS surfaces and solids. There are two kinds of AutoCAD solids: native solids and parametric solids. You can construct a native solid with basic AutoCAD commands. For parametric solids, you need to use AutoCAD Designer commands. AutoCAD Designer is the counterpart of AutoSurf in the Mechanical Desktop Package.

In this chapter, you will build the native solid model together with the surface model of the bottom piece for the mobile phone that you created in the last chapter. You will begin by constructing a native solid model. Then, you will make a NURBS surface to cut the solid. After cutting, you will continue to work on the solid by adding features to it. Finally, you will convert the solid model to a set of surfaces.

With this surface model and the one created in the last chapter, you will learn how to make an assembly drawing using external referencing.

Figure 5.1 shows the native solid model of the bottom piece for the mobile phone. Figure 5.2 is the set of surfaces converted from the solid model. The surface model is a set of surface entities. The solid model is a single, solid entity.

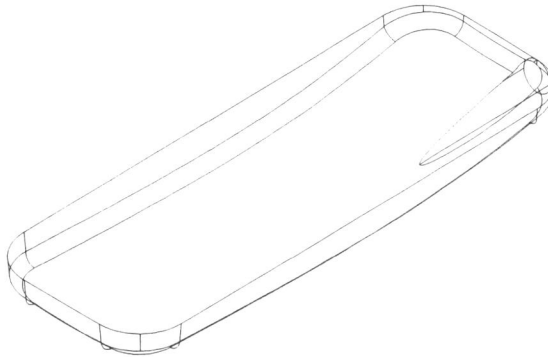

Figure 5.1 Native solid model for the bottom piece of the mobile phone

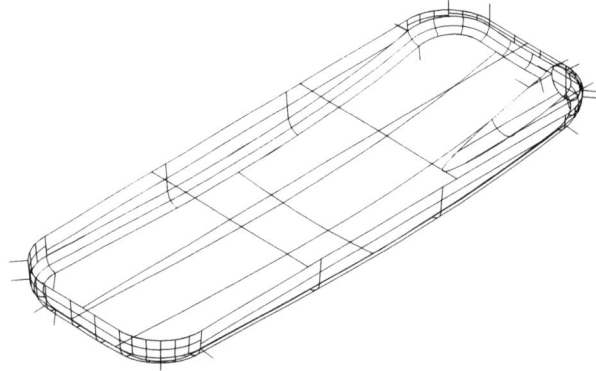

Figure 5.2 Set of surfaces converted from the native solid

5.1 Inter-Operation of Native Solid and NURBS Surfaces

In the last chapter, you created the top piece of the mobile phone. In order to maintain consistent dimensions, you should use the drawing file that you saved in the last chapter as the prototype drawing. Then, you will erase all the irrelevant objects and keep only the necessary wireframes. From the wireframes, you will start to build the native solid model.

Apply the NEW command. Specify the filename of the saved drawing of the mobile phone that you created in the last chapter as the prototype drawing.

    <File>   **<New...>**

Command: **NEW**

        [Prototype: **The saved drawing of the mobile phone**]

Set the current layer to SURF, turn on all the layers, and then turn off the layer WIRE.

    <Data>   **<Layers...>**

Command: **DDLMODES**

Layer	
WIRE	**Off**
SURF	**On**
UTY	**On**

Current layer: **SURF**

With the layer WIRE turned off, you can safely erase all the irrelevant surface entities without worrying about accidentally erasing the required entities.

    [Modify]   **[Erase]**

Command: **ERASE**
Select objects: **[Select all the surfaces.]**
Select objects: **[Enter]**

Turn on the layer WIRE, and set it as the current layer. Figure 5.3 shows the wireframes created on this layer.

<Data> **<Layers...>**

Command: **DDLMODES**

Layer	
WIRE	**On**

Current layer: **WIRE**

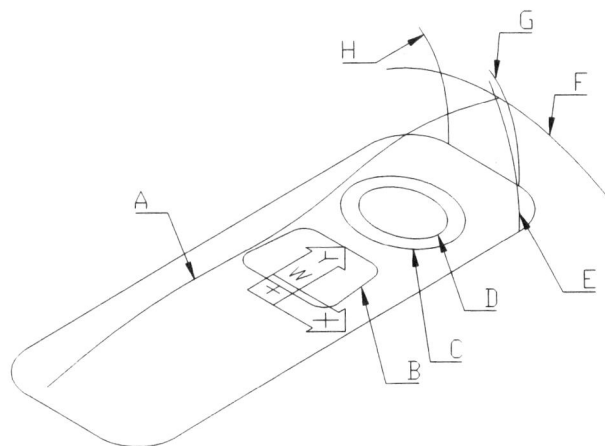

Figure 5.3 Entities residing on layer WIRE

You will need only the base wireframes for this model. Use the ERASE command to erase the irrelevant wireframes. See Figure 5.4.

[Modify] **[Erase]**

Command: **ERASE**
Select objects: **[Select A, B, C, D, E, F, G, and H (Figure 5.3).]**
Select objects: **[Enter]**

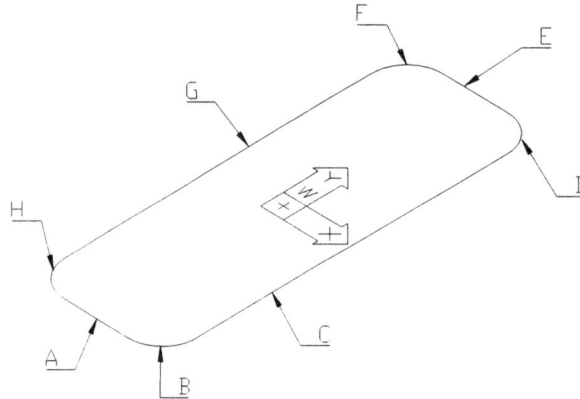

Figure 5.4 Irrelevant entities erased

Run the PEDIT command to join the entities as a polyline.

[Modify]   **[Edit Polyline]**

Command: **PEDIT**
Select polyline: **[Select A (Figure 5.4).]**
Object selected is not a polyline
Do you want to turn it into one? <Y> **Y**
Close/Join/Width/Edit vertex/Fit/Spline/Decurve/Ltype gen/Undo/eXit <X>: **J**
Select objects: **[Select B, C, D, E, F, G, and H (Figure 5.4).]**
Select objects: **[Enter]**
7 segments added to polyline
Open/Join/Width/Edit vertex/Fit/Spline/Decurve/Ltype gen/Undo/eXit <X>: **X**

Set the Z axis of the UCS to align with the current X axis by using the UCS command.

[UCS]   **[Z Axis Vector UCS]**

Command: **UCS**
Origin/ZAxis/3point/OBject/View/X/Y/Z/Prev/Restore/Save/Del/?/<World>: **ZA**
Origin point: **0,0,0**
Point on positive portion of Z-axis: **1,0**

To create a set of wireframes for defining the contour of the base, run the SPLINE command to draw a spline. See Figure 5.5.

[Polyline]   **[Spline]**

Command: **SPLINE**
Object/<Enter first point>: **80,0**
Enter point: **60,-5**
Close/Fit Tolerance/<Enter point>: **20,-10**
Close/Fit Tolerance/<Enter point>: **-20,-10**
Close/Fit Tolerance/<Enter point>: **-60,-8**
Close/Fit Tolerance/<Enter point>: **-80,-7**

Close/Fit Tolerance/<Enter point>: **[Enter]**
Enter start tangent: **[Enter]**
Enter end tangent: **[Enter]**

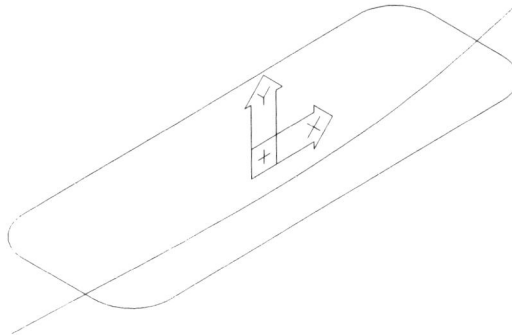

Figure 5.5 Spline created

Use the COPY command to make two copies of the spline. See Figure 5.6.

[Modify]   **[Copy Object]**

Command: **COPY**
Select objects: **LAST**
Select objects: **[Enter]**
<Base point or displacement>/Multiple: **0,2,30**
Second point of displacement: **[Enter]**

[Modify]   **[Copy Object]**

Command: **COPY**
Select objects: **LAST**
Select objects: **[Enter]**
<Base point or displacement>/Multiple: **0,0,-60**
Second point of displacement: **[Enter]**

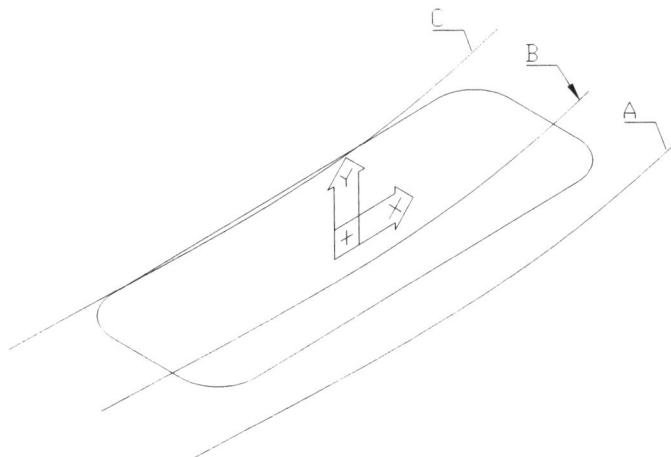

Figure 5.6 Two splines copied

You have completed all the necessary wireframes for this model. Create a new layer called SOLID and set the current layer to SURF.

<Data>   **<Layers...>**

Command: **DDLMODES**

Layer	Color
SOLID	**green**

Current layer: **SURF**

You will create a NURBS surface. Before doing so, use the AMSURFVARS command to set a few AutoSurf system variables.

<Surfaces> **<Preferences...>**

Command: **AMSURFVARS**

  [Surface Properties
  U Display Wires **2**
  V Display Wires **1**
  Vector Length **4**
  Keep original **Yes**
  OK   ]

Run the AMLOFTU command to create a lofted U surface from the three splines. See Figure 5.7.

<Surfaces> **<Create Surface>** **<Loft U...>**

Command: **AMLOFTU**
Select U wires: **[Select A, B, and C (Figure 5.6).]**
Select U wires: **[Enter]**

  [Input Wires
  Curve Direction **Align**
  Curve Fit   **Smooth**
  OK   ]

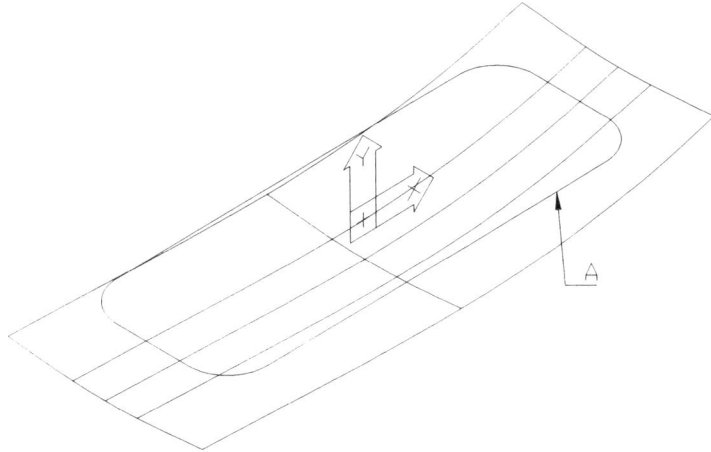

Figure 5.7 Lofted U surface created

You will use this lofted U surface to cut a solid. To cut a solid, the surface must be large enough to extend beyond the boundary of the solid.

Set the current layer to SOLID. Then, run the EXTRUDE command to create a native solid by extruding the polyline. See Figure 5.8.

<Data>   **<Layers...>**

Command: **DDLMODES**
Current layer: **SOLID**

[Solids]   **[Extrude]**

Command: **EXTRUDE**
Select objects: **[Select A (Figure 5.7).]**
Select objects: **[Enter]**
Path/<Height of Extrusion>: **-20**
Extrusion taper angle: **5**

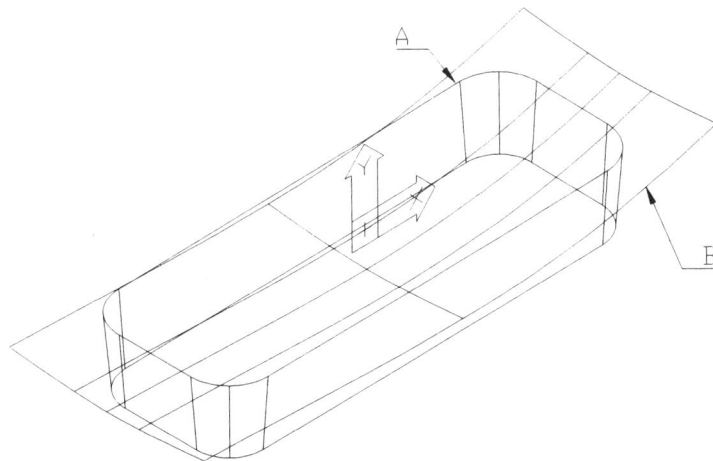

Figure 5.8 Extruded solid created

The limitation of using only a solid in model creation is the absence of free-form surface features. You can overcome this limitation by using AutoSurf. Run the AMSOLCUT command to cut the native solid with the surface.

<Surfaces>   <Edit Solid>

Command: **AMSOLCUT**
Select solid to cut: **[Select A (Figure 5.8).]**
Select surface: **[Select B (Figure 5.8).]**
Portion to remove: Flip/<Accept>: **A**   **[Examine the normal vector direction. If it is pointing upward, you should choose FLIP and then ACCEPT.]**

You can use an AutoSurf surface to cut a native solid or a parametric solid. The AMSOLCUT command works on native solids. If you want to cut a parametric solid created with AutoCAD Designer, you need to use the AMSURFCUT command — an AutoCAD Designer command.

After cutting the solid, use the AMVISIBLE command to hide the surface. Then, turn off the layers SURF and WIRE. See Figure 5.9.

<Surfaces>   <Object Visibility...>

Command: **AMVISIBLE**

        **[Hide
          Select   ]**

Select objects to hide: **[Select B (Figure 5.8).]**
Select objects to hide: **[Enter]**

        [OK]

<Data>   <Layers...>

Command: **DDLMODES**

Layer	
SURF	**Off**
WIRE	**Off**

Current layer: **SOLID**

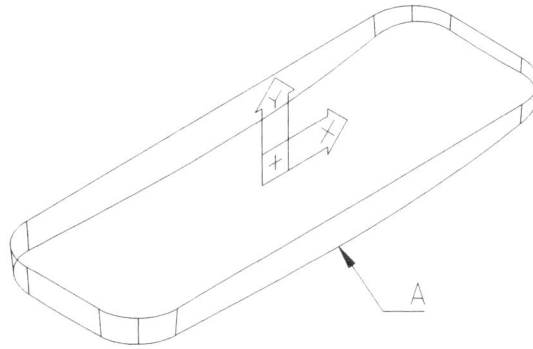

Figure 5.9 Solid cut by a surface

Solid model creation does not end after you cut a solid with the AMSOLCUT command. To continue working on the solid, run the FILLET command to round the edges of the solid. See Figure 5.10.

[**Modify**]   [**Fillet**]

Command: **FILLET**
(TRIM mode) Current fillet radius = 0.0000
Polyline/Radius/Trim/<Select first object>: **[Select A (Figure 5.9).]**
Enter radius <7.0000>: **3**
Chain/Radius/<Select edge>: **C**
Edge/Radius/<Select edge chain>: **[Select A (Figure 5.9).]**
Edge/Radius/<Select edge chain>: **[Enter]**

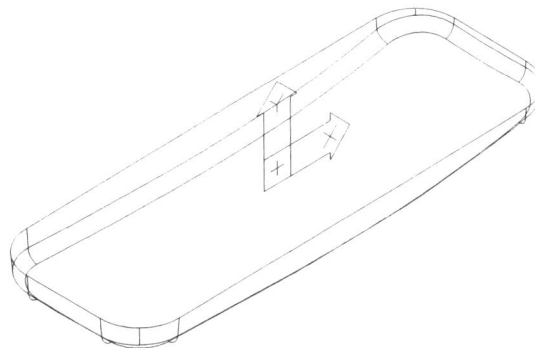

Figure 5.10 Edges filleted

Align the Z axis of the UCS with the negative X direction by using the UCS command. Then, construct a solid cylinder by using the CYLINDER command. See Figure 5.11.

[**UCS**]   [**Z Axis Vector UCS**]

Command: **UCS**

Origin/ZAxis/3point/OBject/View/X/Y/Z/Prev/Restore/Save/Del/?/<World>: **ZA**
Origin point: **0,0,0**
Point on positive portion of Z-axis: **-1,0**

[Solids]   **[Cylinder]**   **[Center]**

Command: **CYLINDER**
Elliptical/<center point>: **10,-5**
Diameter/<Radius>: **4**
Center of other end/<Height>: **-66**

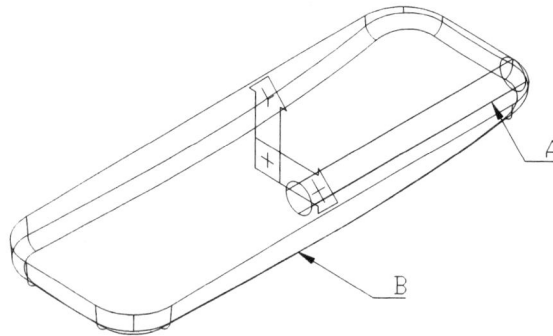

Figure 5.11 Solid cylinder created

To reiterate, you can continue to modify a solid after you have cut it with a surface. Use the UNION command to unite the two solids. See Figure 5.12.

Command: **UNION**
Select objects: **[Select A (Figure 5.11).]**
Select objects: **[Select B (Figure 5.11).]**
Select objects: **[Enter]**

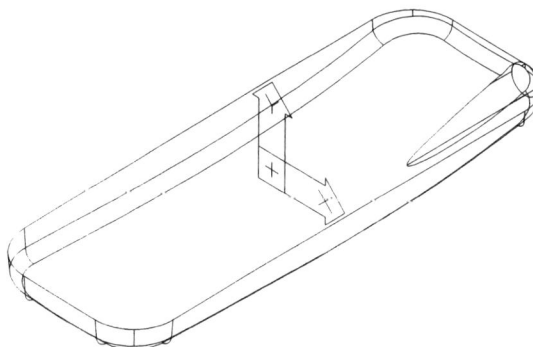

Figure 5.12 Two solids united

The solid model for the bottom piece of the mobile phone is complete. By now, you should know how to use a NURBS surface to cut a native solid and continue to work on the solid.

The opposite direction of inter-operation of AutoSurf with a native solid is to convert a native solid to a set of NURBS surfaces. There are times when you will want to convert a solid model to a surface model. One good reason might be that you want a variable fillet, which cannot be done in a native solid or a parametric solid.

To appreciate how a solid model can be converted to a set of surfaces, you will convert the solid model of the bottom piece of the mobile phone to a set of surfaces.

Set the current layer to SURF. Then, run the AM2SF command. See Figure 5.13.

<Data>   **<Layers...>**

Command: **DDLMODES**
Current layer: **SURF**

<Surfaces>   **<Create Surface>**   **<From ACAD>**

Command: **AM2SF**
Face/<Objects>: **OBJECTS**
Select objects: **[Select the solid model.]**
Select objects: **[Enter]**
Converting: 1 of 1 .....

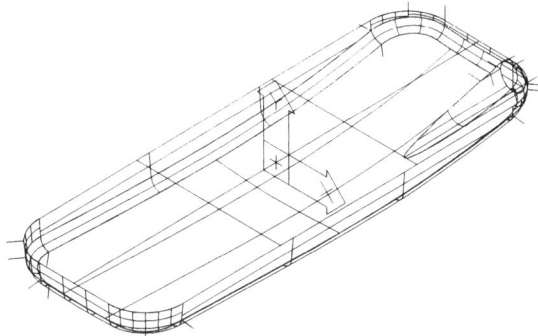

Figure 5.13 Complex solid converted to a set of surfaces

Now, you have a set of surfaces in addition to a solid model. The set of surfaces resides on the layer SURF, and the solid model resides on the layer SOLID. Turn off the layer SOLID. Then, set the UCS to WORLD. See Figure 5.2 again.

<Data>   **<Layers...>**

Command: **DDLMODES**

Layer	
SOLID	**Off**

Current layer: **SURF**

[UCS]   **[World UCS]**

Command: **UCS**
Origin/ZAxis/3point/OBject/View/X/Y/Z/Prev/Restore/Save/Del/?/<World>: **W**

Run the **QSAVE** command to save your work to a file.

<File>   <Save...>

Command: **QSAVE**

5.2   Assembly of Surface Models

After the completion of the surface model of the base for the mobile phone, you will start
a new drawing, and import the two surface models to create an assembly.
Run the NEW command.

<File>   <New...>

Command: **NEW**

Set the display to an isometric view with the accclerator key [8].

Command: **8**

Import the upper casing of the mobile phone surface model by using the XREF
command. See Figure 5.14

[External Reference]   [Attach]

Command: **XREF**
?/Bind/Detach/Path/Reload/Overlay/<Attach>: **A**
Attach Xref AS04R3C: **[Specify the filename for the mobile phone top.]**
Insertion point: **0,0**
X scale factor <1> / Corner / XYZ: **1**
Y scale factor (default=X): **[Enter]**
Rotation angle <0>: **[Enter]**

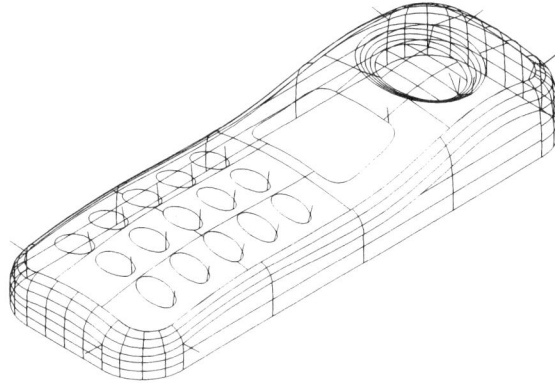

Figure 5.14 Mobile phone top body

Repeat the XREF command to import the lower casing of the mobile phone. See Figure 5.15.

[**External Reference**]   [**Attach**]

Command: **XREF**
?/Bind/Detach/Path/Reload/Overlay/<Attach>: **A**
Attach Xref AS04R3C: **[Specify the filename for the mobile phone base.]**
Insertion point: **0,0**
X scale factor <1> / Corner / XYZ: **1**
Y scale factor (default=X): **[Enter]**
Rotation angle <0>: **[Enter]**

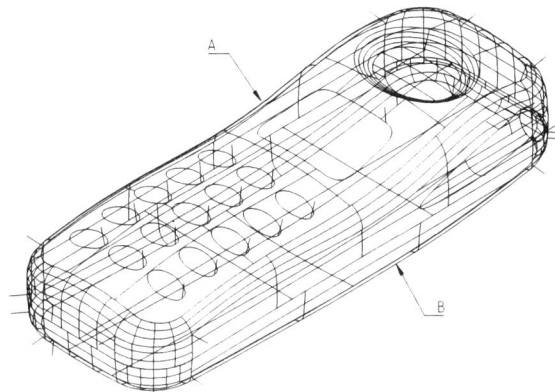

Figure 5.15 Mobile phone base

To rotate the two objects 90° about the current X axis, use the ROTATE3D command. See Figure 5.16.

[**Modify**]   [**3D Rotate**]

Command: **ROTATE3D**
Select objects: **[Select A and B (Figure 5.15).]**

```
Select objects: [Enter]
Axis by Object/Last/View/Xaxis/Yaxis/Zaxis/<2points>: X
Point on X axis <0,0,0>: [Enter]
<Rotation angle>/Reference: 90
```

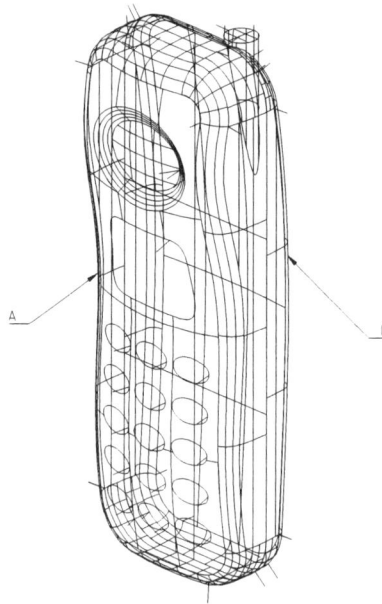

Figure 5.16 Objects rotated

The assembly is complete.

External referencing has two advantages over insertion: The file size of the assembly drawing is small, and the drawing always shows the most updated version of the external drawings. However, there is a drawback. You cannot modify the externally referenced object in the assembly drawing.

If you want to edit the surface models individually, you must go back to the source drawing to which the current drawing is externally referenced and do the necessary editing. If you want to edit in the current drawing, you have to import the data of the source drawing to become permanent data of this drawing.

To permanently import the data of the source drawing, run the XREF command with the BIND option to bind the external objects into the current drawing. After binding, the two models become part of the current drawing, and you can edit the objects in the assembly. However, any future updates to the source drawings in the future will not show up here.

```
Command: XREF
?/Bind/Detach/Path/Reload/Overlay/<Attach>: B
Xref(s) to bind: *
```

Before you can work on the bound objects, you have to explode them because they have become instances of the internal blocks after binding. Run the EXPLODE command.

[Modify]   **[Explode]**

Command: **EXPLODE**
Select objects: **[Select A and B (Figure 5.16).]**
Select objects: **[Enter]**

Because there is no noticeable change on your screen after using the EXPLODE command, it is very likely that you might run the command more than once on an object. Be warned, however, that each time the EXPLODE command is used on an object, the object decreases in complexity by one level. If you overexplode an object, you might end up with a cluster of lines and arcs.

After binding and exploding, use the AMDISPSF command to set the normal vector length, the number of U-lines, and the number of V-lines for the two sets of surfaces.

<Surfaces>   **<Display...>**

Command: **AMDISPSF**
Select surfaces: **[Select all the surfaces.]**
Select surfaces: **[Enter]**

[Persistent Display:
Normal Length: **5**
Number of U-lines: **2**
Number of V-lines: **1**
OK   ]

You have completed the assembly of two surface models. Use the accelerator key [4] to set the display to four viewports. See Figure 5.17.

Command: **4**

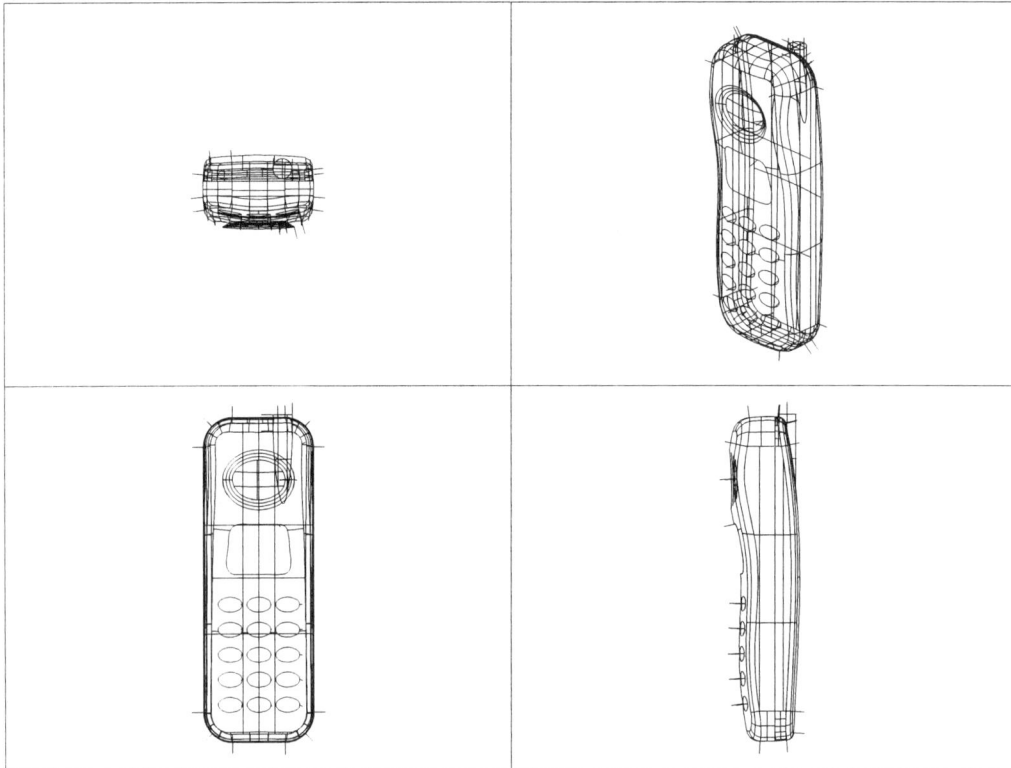

Figure 5.17 Assembly in four viewports

Save the drawing with the SAVE command.

Command: **SAVE**

5.3   Summary

In this chapter, you have practiced the AMSOLCUT command, in addition to those commands that you learned in the previous chapters. For a brief explanation of this command, refer to the appendix of this book.

In completing the project, you practiced how to use a surface as a cutting plane to cut a solid, and then to continue on working with the cut solid. You also converted the solid to a set of AutoSurf surfaces.

As mentioned earlier in this chapter, AutoSurf is a part of the Mechanical Desktop Package, and is fully compatible with AutoCAD Designer and AutoCAD native solids. To create a complex design, you can use AutoSurf in conjunction with the other two applications.

AutoSurf is useful for making free-form NURBS surfaces. Depending on your design intent and design criteria, you can use AutoSurf alone to create a model if you need only surfaces. However, it might be more convenient and faster to convert some existing entities that you created in AutoCAD to surfaces. In Chapters 2 and 3, you created a circle with thickness, which is a typical 2D AutoCAD object, and converted it to a cylindrical surface. In Chapter 3, you also created two solid boxes, which are native solids, and converted them to twelve pieces of planar surfaces.

The other two counterparts of AutoSurf, native solids and parametric solids, are useful solid modeling tools. Native solid is straightforward and simple, but cannot be edited. Therefore, you might use native solids in one situation and parametric solids in another. The limitation of using the solid modelers alone is that free-form surface features are not readily available. Therefore, you will need to use AutoSurf to provide the necessary free-form feature to combine with a solid.

Basically, you can use any kind of NURBS surfaces to cut a solid. The only thing you need to remember is that the surface must be large enough to extend beyond the boundary of the solid that you are cutting.

To sum up, the three modules can inter-operate, as illustrated in Figure 5.18. A NURBS surface can be used to cut a native solid or a parametric solid. To cut a native solid, you can use the AMSOLCUT command in AutoSurf. To cut a parametric solid, you can use the AMSURFCUT command in AutoCAD Designer. To convert either a native solid or a parametric solid to a set of NURBS surfaces, you can use the AM2SF command in AutoSurf. Regarding the other two solids, you can convert a parametric solid to a native solid by using the EXPLODE command, and convert a native solid to a static base solid feature for subsequent addition of the parametric solid features by using the AMNEWPART command, which is part of AutoCAD Designer.

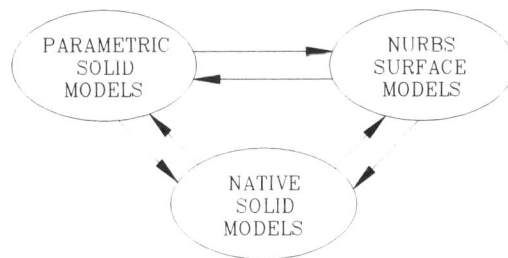

Figure 5.18 Inter-Operation

Besides inter-operation, you can also use the external referencing facilities to assemble two sets of NURBS surfaces. Remember, you must be very careful using the EXPLODE command. Each time you use this command on an object, you break the object down one level of complexity. If you explode a block, you get the original entities. If you explode further, you might break the entities into their more primitive forms, such as lines, arcs, and so forth.

In the next chapter, you will learn how to prepare a 2D drawing document from a surface model.

5.4   Exercise

To further enhance your knowledge of inter-operation of solid models and free-form surfaces, you will work on the model of an engine connecting rod. See Figure 5.19.

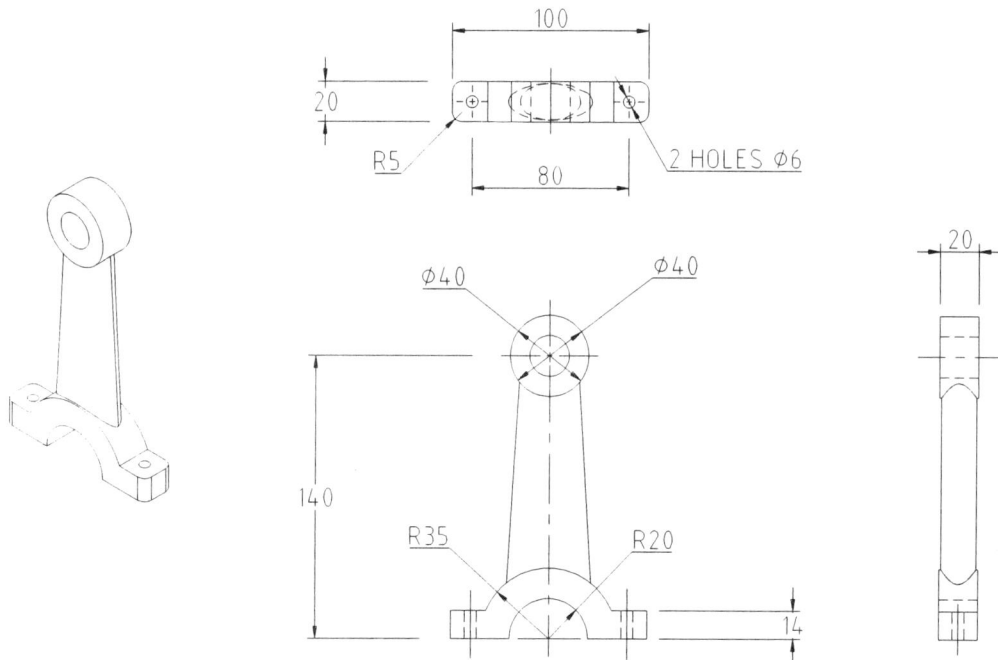

Figure 5.19 Solid model of an engine connecting rod

The central portion of the connecting rod is elliptical with a cross-section that varies linearly in size along its axis. To make this portion of the model, you should create a solid box and a ruled surface. Then, you will use the ruled surface to cut the solid box. To create the solid model, you can use either a native solid or a parametric solid. If you use a native solid, you should use the AMSOLCUT command to cut the solid box. If you use a parametric solid, you should use the AMSURFCUT command to cut the solid box.

Figure 5.20 shows the wireframes for the two ends of the connecting rod. In making the wireframes, set the origin at the center of the lower rod ends.

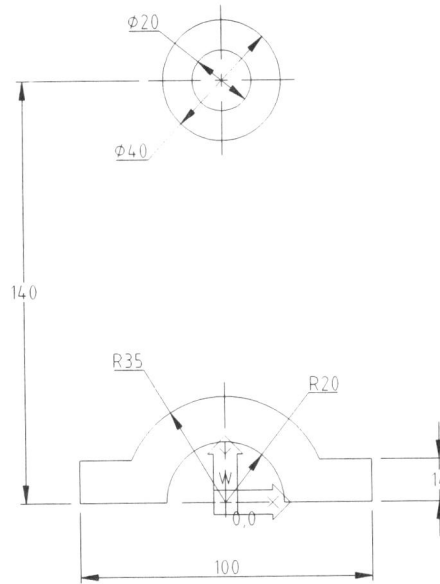

Figure 5.20 Wireframes for making the connecting rod ends

As shown in Figure 5.21, create two ellipses. These ellipses reside on the ZX plane of WCS.

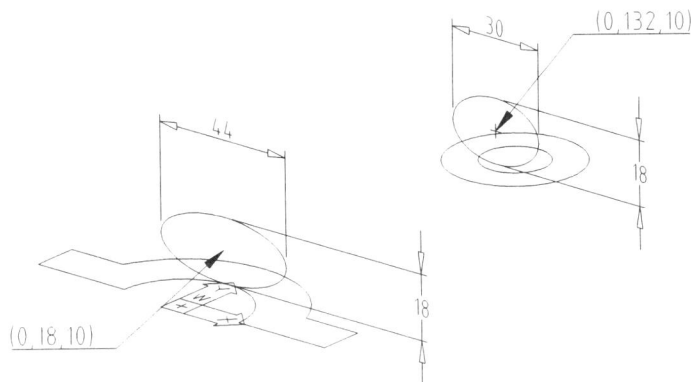

Figure 5.21 Two ellipses for making the ruled surface

Make a solid box. See Figure 5.22.

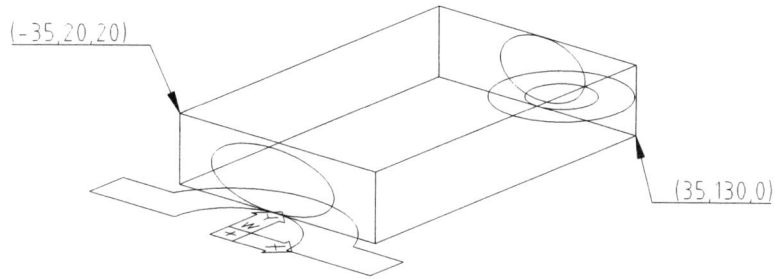

Figure 5.22 Solid box created

Using the two ellipses as input wires, create a ruled surface. See Figure 5.23.

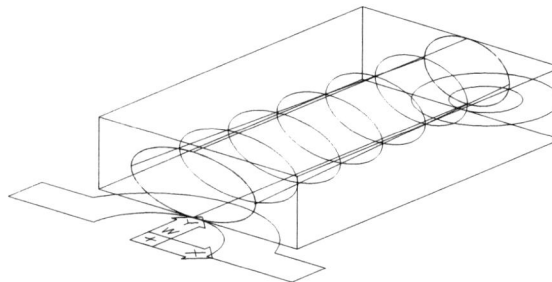

Figure 5.23 Ruled surface created

Use the ruled surface to cut the solid box. Then, extrude the two ends of the connecting rods. See Figure 5.24.

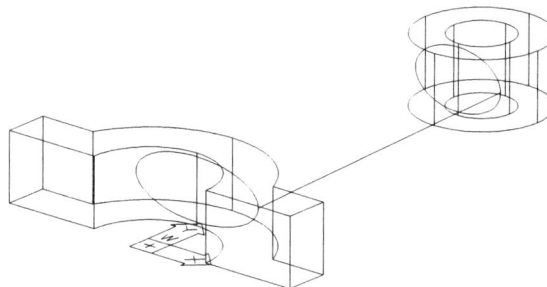

Figure 5.24 Box cut by ruled surface and two ends extruded

Join the central portion and the two ends. See Figure 5.25.

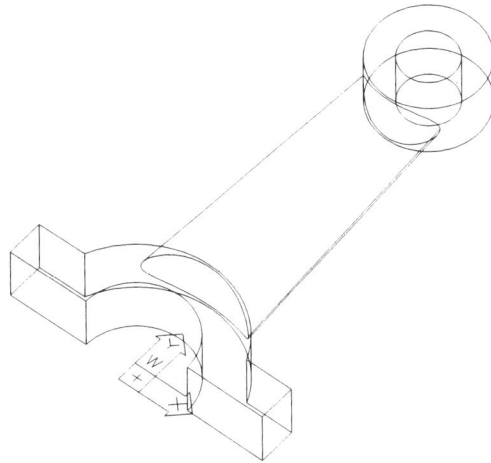

Figure 5.25 Two ends joined with the central part

To complete the solid model, add two holes and fillet four edges. See Figure 5.26.

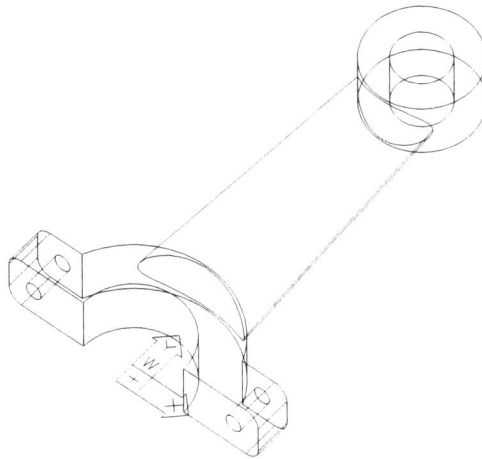

Figure 5.26 Two holes added and four corners filleted

Save your drawing.

Chapter 6
Documentation

A NURBS surface model can serve many purposes. You can use it to manufacture the component described by the model or to create a photo-realistic rendering. Sometimes, you simply might want to create a 2D engineering drawing from it.

This chapter will guide you to produce an orthographic engineering drawing from the surface model of the bottom piece of the mobile phone. You will start by setting up a proper engineering drawing title block in paper space. Then, you will create orthographic viewports using AutoSurf commands. Figure 6.1 shows the completed engineering drawing.

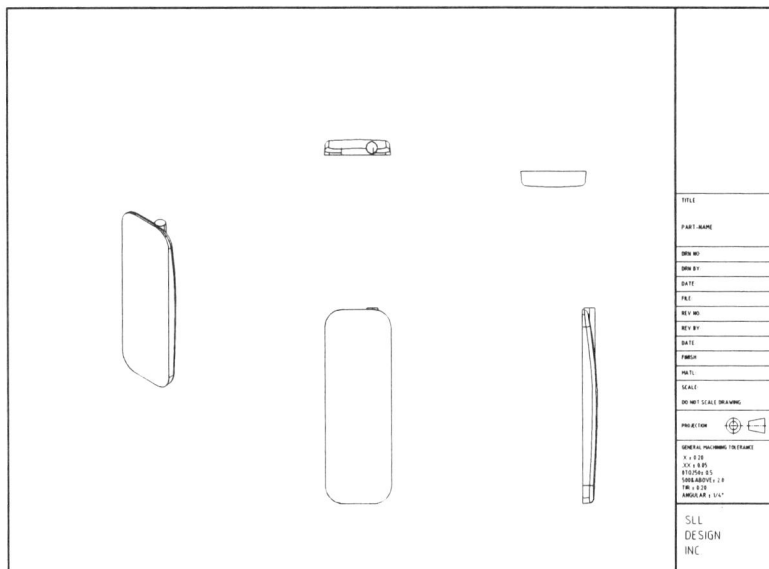

Figure 6.1 Engineering drawing of the mobile phone bottom piece

6.1 Setting Up a Drawing Title Block

As a standard practice, you need to have a title block in your drawing document. Usually, the title block includes a company logo, information about the drawing, and important notes. Figure 6.2 is a sample drawing of a Letter size (or A4) title block. If you do not have a title block, you should create one before proceeding.

Figure 6.2 Typical Letter size (A4) title block

Open the drawing of the mobile phone base piece that you created in the last chapter. See Figure 6.3.

<File>   **<Open...>**

Command: **OPEN**
File name: **[Enter the filename of the base of the mobile phone.]**

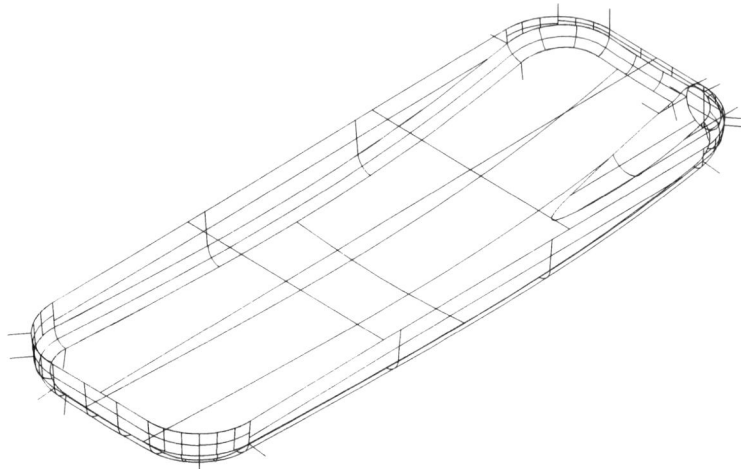

Figure 6.3 Surface model in model mode

AutoCAD has two working environments in which you can place entities: the model space and the paper space. The model space is the environment in which you create the main component of the drawing — the surface model. The paper space is the place where you prepare an engineering drawing document from the surface model. To toggle between the two spaces, you manipulate the system variable TILEMODE. The default value of this variable is one, the model space. Setting it to zero switches the environment to paper space.

In AutoSurf, the two working spaces have different names: the drawing mode and the model mode. The drawing mode is the paper-space environment, and the model mode is the model-space environment. You can toggle between drawing mode and model mode with the AMMODE command.

The starting point of document preparation is to switch to drawing mode and to insert a title block.

Run the AMMODE command.

<Drawings>   **<Drawing Mode>**

Command: **AMMODE**
Model/<Drawing>: **[Enter]**

When you switch to drawing mode the first time, the screen becomes blank. Temporarily, you cannot see the surface model that resides in model mode. If you have turned on the UCS icon, the icon changes to the shape of a set square.

Make a new layer TITLE. You will place the title block on this layer.

<Data>   **<Layers...>**

Command: **DDLMODES**

Layer
TITLE

Current layer: **TITLE**

Use the INSERT command to insert the title block drawing at the (0,0) position with a scale factor of one for both the X axis and Y axis. The rotation angle of insertion should be zero. Using an insertion scale of one and rotation angle of zero simplifies subsequent plot setup. When you plot, you can simply plot the drawing with a scale of one to one at the plot origin of (0,0).

Command: **INSERT**
Block name (or ?): **[Enter the filename of the title block.]**
Insertion point: **0,0**
X scale factor <1>/Corner/XYZ: **1**
Y scale factor (default=X): **1**
Rotation angle <0>: **0**

Now, you have a title block in drawing mode and your surface model in model mode. Your screen should look like Figure 6.2.

6.2   Creating Views in Drawing Mode

After inserting a title block in drawing mode, you will create the engineering drawing views by using the AMDWGVIEW command. This command is the same in both AutoSurf and AutoCAD Designer. However, some options do not apply to AutoSurf surfaces.

Before you start to make the drawing views, you have to decide the following:

Angle of projection (First/Third)
Drafting standard (ANSI/DIN/ISO/JIS)
Centerline geometry and linetype
Hidden line linetype

Upon decided the drawing standards, you should use the AMDWGVARS command to adjust the system variables related to engineering drawing creation. In this drawing, you will use the third angle projection and the ISO drawing standard. See Figure 6.4.

<Drawings>   **<Preferences...>**

Command: **AMDWGVARS**

[Projection type of unfolded views: **Third Angle**
 Display parametric dimensions (applicable to AutoCAD Designer)
 Hide drawing viewport borders   **Yes**
 Drafting Standard
          ISO for all
          Centerlines
                   Centermark   **2**
                   Gap   **2**
                   Overshoot   **3**
                   Centerline linetype   **CENTER**
 Section symbol linetype: (applicable to AutoCAD Designer)
 Hidden line linetype: **HIDDEN**
 OK   ]

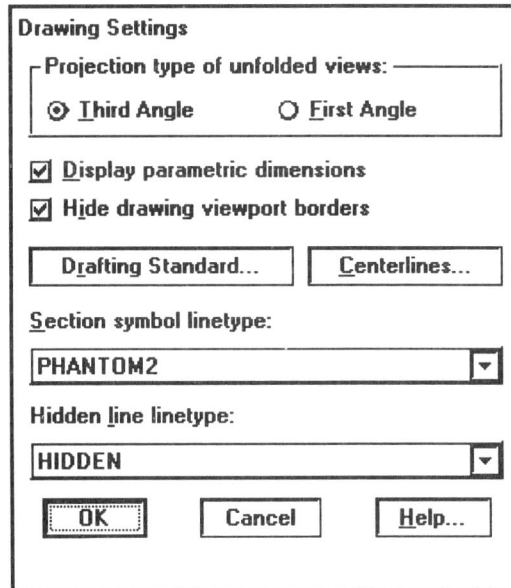

Figure 6.4 Drawing Settings dialog box

To make a set of engineering drawing views of a surface model in drawing mode, you will use the AMDWGVIEW command.

As mentioned above, this command is common to AutoSurf and AutoCAD Designer. To make a view, it requires you to decide which entities to display in terms of a data set. There are four data set types: Active part, Assembly, Select and Group. Active part applies to parametric solid parts created with AutoCAD Designer. Assembly applies to the assembly of solid parts or parametric solid parts created with AutoCAD Designer. For AutoSurf objects, you can choose only between Select or Group. If you use the Select option, you have to select the surface objects to include in the drawing views. If you use the Group option, you have to predefine a group of entities in the model space, and then use the group in the drawing view.

Because a surface model is a set of individual surface entities, to select them one by one while creating the drawing views is very inconvenient. Thus, the Group option is the best option for surface models.

To use the Group option, you have to switch back to model mode with the AMMODE command, and then use the GROUP command to put all the surfaces in an entity group.

<**Drawings**>   <**Drawing Mode**>

Command: **AMMODE**
Drawing/<Model>: **[Enter]**

[Standard Toolbar]   **[Object Group]**

Command: **GROUP**

    [Group name: **PHONE**
    Selectable: **Yes**
    New:]

Select objects for grouping:
Select objects: **[Select all the surfaces.]**
Select objects: **[Enter]**

After grouping, switch to drawing mode again.

<Drawings>   **<Drawing Mode>**

Command: **AMMODE**
Model/<Drawing>: **[Enter]**

The first drawing view in a document has to be a base view. To speed up drawing creation, you should hide the hidden lines and the tangencies. The drawing view is a floating viewport. It is an entity and a border. Because you have already set Hide drawing viewport borders to Yes when you ran the AMDWGVARS command, the border of the drawing views will be hidden.

Run the AMDWGVIEW command to create the base view. See Figure 6.6.

<Drawings>   **<Create View>**

Command: **AMDWGVIEW**

       [Type: **Base**
        Hidden Lines:
                Calculate hidden lines   **Yes**
                Hide hidden lines   **Yes**
                Linetype of hidden lines: **HIDDEN**
                Display tangencies   **No**
        Data Set: **Group**
        Scale: **0.5**
        OK   ]

Enter group name to be included in view: **PHONE**
worldXy/worldYz/worldZx/Ucs/<Select work plane or edge>: **WORLDXY**
worldX/worldY/worldZ/<Select work axis or straight edge>: **WORLDX**
Rotate/Z-flip/<Accept>: **[Enter]**
Location for base view: **[Select A (Figure 6.2).]**
Location for base view: **[Enter]**

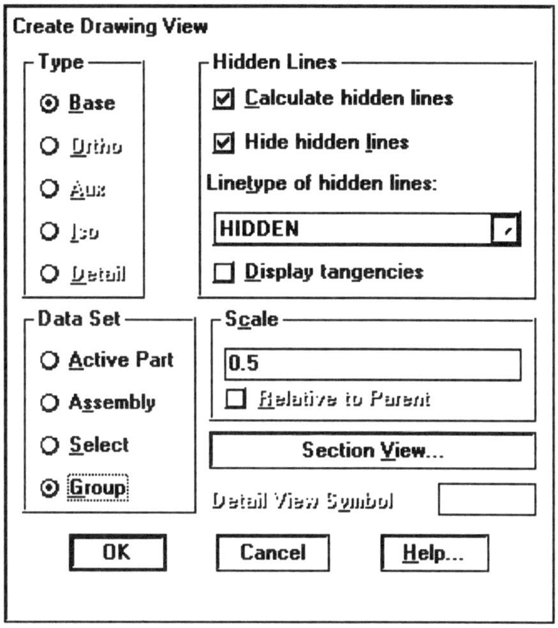

Figure 6.5 Create Drawing View dialog box

Figure 6.6 shows the base view of the drawing.

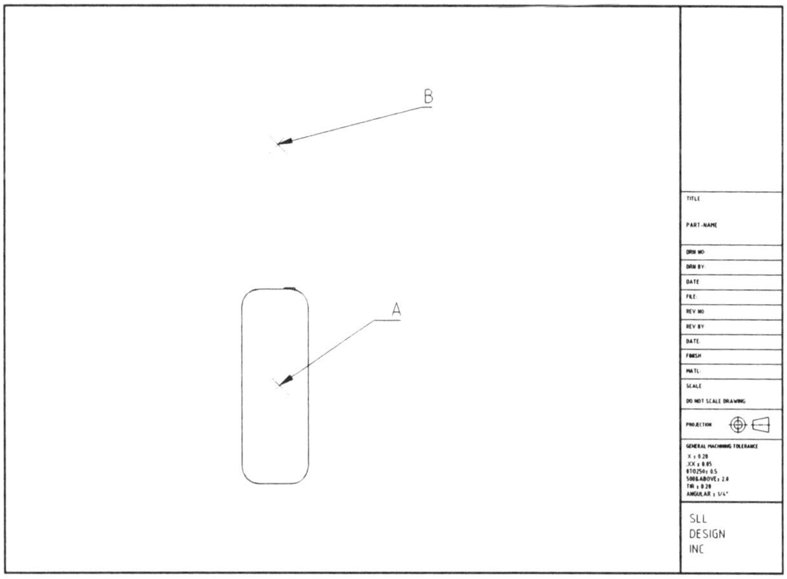

Figure 6.6 Base view created

After making the first base view in a drawing, the second drawing view can be one of the following types: Base, Ortho (orthographic), Aux (auxiliary), Iso (isometric), or Detail (enlarged partial view). Except for the base view, all views need a parent view.

You can set up another drawing view that is totally unrelated to the existing drawing view if you choose the Base view again. Very seldom will you make more than one base view in a drawing because it is not standard practice to create unrelated views in a drawing.

Treating the drawing already created as the front view of the model, run the AMDWGVIEW command again to make a top view. The parent view of this view is the front view. See Figure 6.7.

<Drawings>   **<Create View>**

Command: **AMDWGVIEW**

[Type: **Ortho**
  Hidden Lines:
        Calculate hidden lines   **Yes**
        Hide hidden lines   **Yes**
        Linetype of hidden lines: **HIDDEN**
        Display tangencies   **No**
    OK   ]

Select parent view: **[Select A (Figure 6.6).]**
Location for orthographic view: **[Select B (Figure 6.6).]**
Location for orthographic view: **[Enter]**

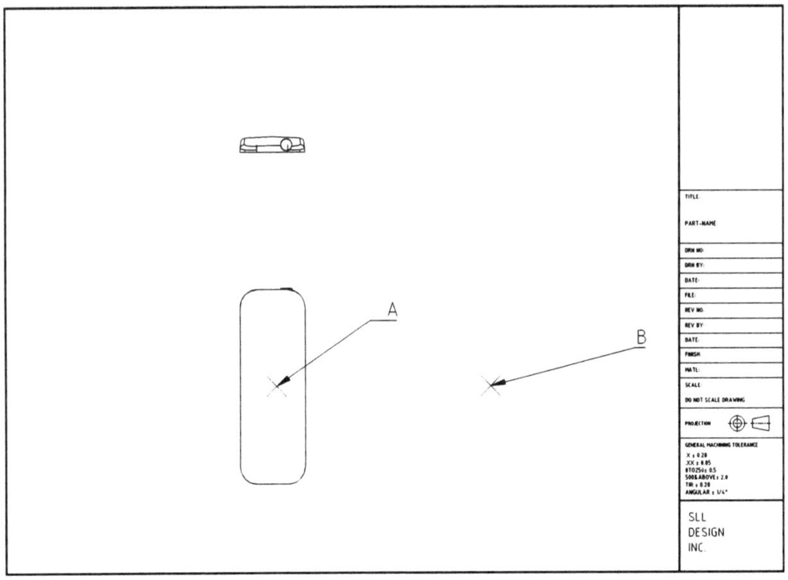

Figure 6.7 Orthographic top view created

Repeat the AMDWGVIEW command to make an orthographic side view at the right side. See Figure 6.8.

<Drawings>   **<Create View>**

Command: **AMDWGVIEW**

[Type: **Ortho**

Hidden Lines:
          Calculate hidden lines   **Yes**
          Hide hidden lines   **Yes**
          Linetype of hidden lines: **HIDDEN**
          Display tangencies   **No**
OK   ]

Select parent view: **[Select A (Figure 6.7).]**
Location for orthographic view: **[Select B (Figure 6.7).]**
Location for orthographic view: **[Enter]**

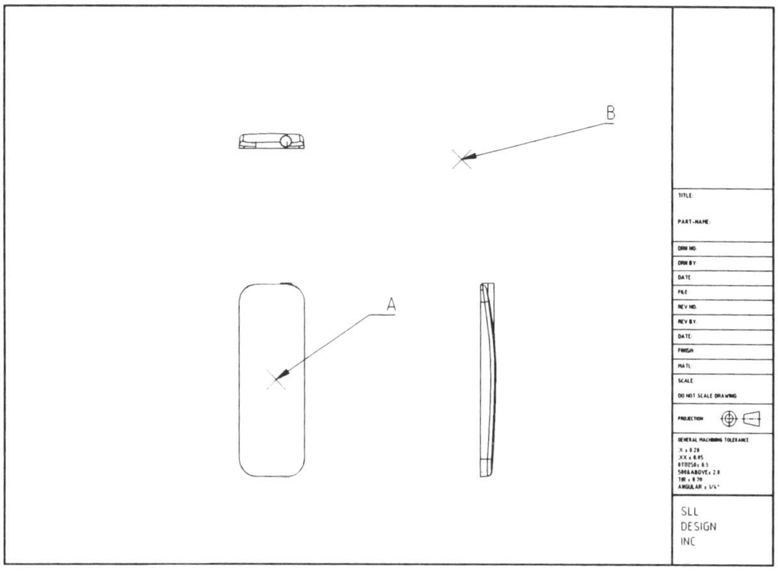

Figure 6.8 Orthographic side view added

After making the front view, top view, and side view, use the AMDWGVIEW
command to create an isometric view. See Figure 6.9. Using any given view as the parent
view, you can create four isometric drawing views — one at the upper-right corner, one
at the upper-left corner, one at the lower-left corner, and one at the lower-right corner.

<Drawings>   <Create View>

Command: **AMDWGVIEW**

[Type: **Iso**
  Hidden Lines:
          Calculate hidden lines   **Yes**
          Hide hidden lines   **Yes**
          Linetype of hidden lines: **HIDDEN**
          Display tangencies   **No**
  Scale: **1**
  Relative to Parent   **Yes**
  OK   ]

Select parent view: **[Select A (Figure 6.8).]**

Location for isometric view: **[Select B (Figure 6.8).]**
Location for isometric view: **[Enter]**

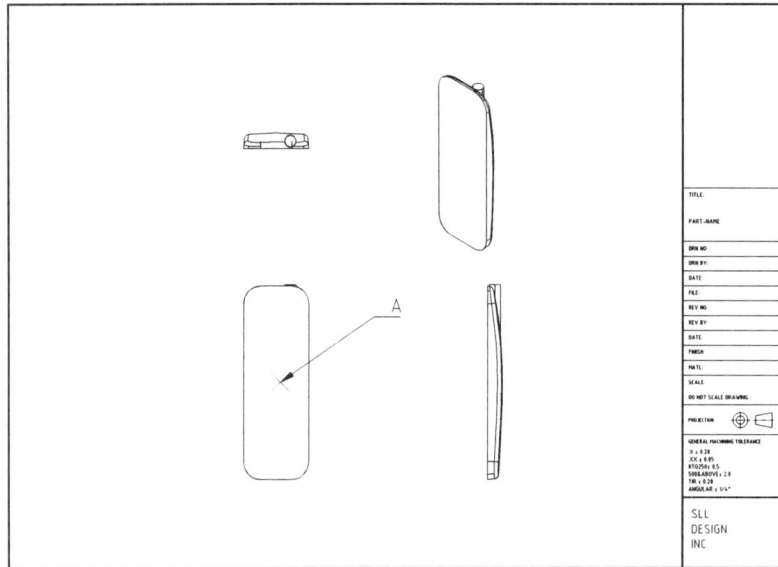

Figure 6.9 Isometric view created

Now, you have an isometric drawing situated at the upper-right corner of the drawing. Later, you will learn how to move it into any position. The next drawing view to create is an auxiliary view. To make an auxiliary view, you will need a straight edge to define the direction of projection. Because there is no straight edge in the surface model, you have to create a straight line as reference. Use the AMMODE command to switch to model mode. Then, draw a line running from (23,65,0) to (-23,-65,20) by using the LINE command. This line defines the viewing direction of the auxiliary view.

<Drawings>   **<Drawing Mode>**

Command: **AMMODE**
Drawing/<Model>: **[Enter]**

[Draw]   **[Line]**

Command: **LINE**
From point: **23,65,0**
To point: **-23,-65,20**
To point: **[Enter]**

After drawing a line, switch back to drawing mode by using the AMMODE command.

<Drawings>   **<Drawing Mode>**

Command: **AMMODE**
Model/<Drawing>: **[Enter]**

Upon switching back to drawing mode, you will find that the line you just created does not appear in the drawing views. To display the line, you have to suppress the calculation of hidden lines. Run the AMEDITVIEW command to edit the front view. See Figure 6.10.

<Drawings>   **<Edit View>**   **<Attributes>**

Command: **AMEDITVIEW**
Select view to edit: **[Select A (Figure 6.9).]**

[Hidden Lines:
  Calculate hidden lines   **No**
  OK   ]

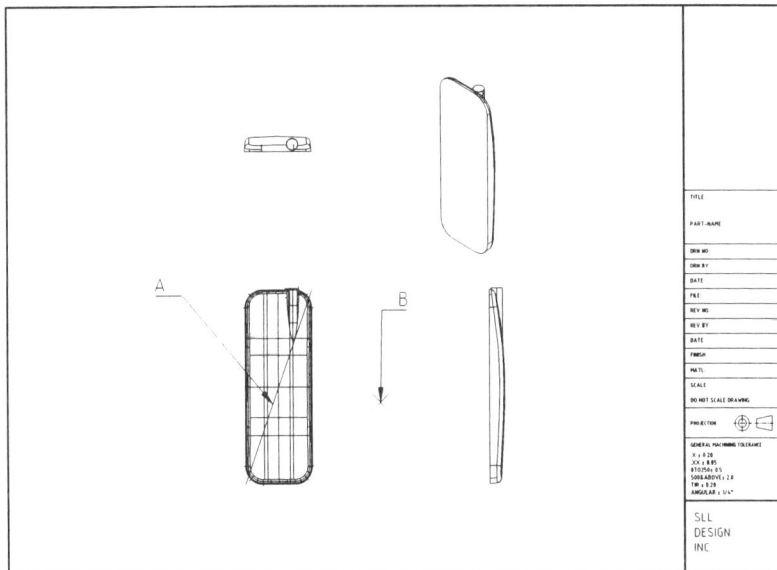

Figure 6.10 Front view edited

As you can see, the line appears in the front view. Now, you can run the AMDWGVIEW command to make an auxiliary view in the direction defined by the line. See Figure 6.11.

<Drawings>   **<Create View>**

Command: **AMDWGVIEW**

[Type: **Aux**
  Hidden Lines:
          Calculate hidden lines   **No**
  OK   ]

Select a straight edge in the parent view: **[Select A (Figure 6.10).]**
Select second point or <RETURN> to use the selected edge: **[Enter]**
Location for auxiliary view: **[Select B (Figure 6.10).]**
Location for auxiliary view: **[Enter]**

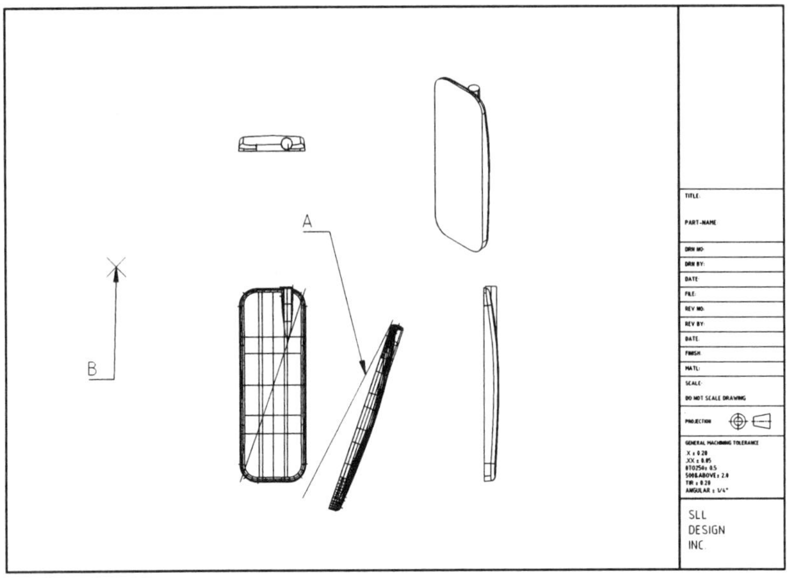

Figure 6.11 Auxiliary view created

In the auxiliary view you just created, the reference line appears because you set Calculate hidden lines for this view to No.

Draw another auxiliary view that uses the previous auxiliary view as the parent view. Run the AMDWGVIEW command. See Figure 6.12. Basically, you can use any drawing view as the parent view.

<Drawings>   **<Create View>**

Command: **AMDWGVIEW**

    [Type: **Aux**
     Hidden Lines:
          Calculate hidden lines   **No**
     OK   ]

Select a straight edge in the parent view: **[Select A (Figure 6.11).]**
Select second point or <RETURN> to use the selected edge: **[Enter]**
Location for auxiliary view: **[Select B (Figure 6.11).]**
Location for auxiliary view: **[Enter]**

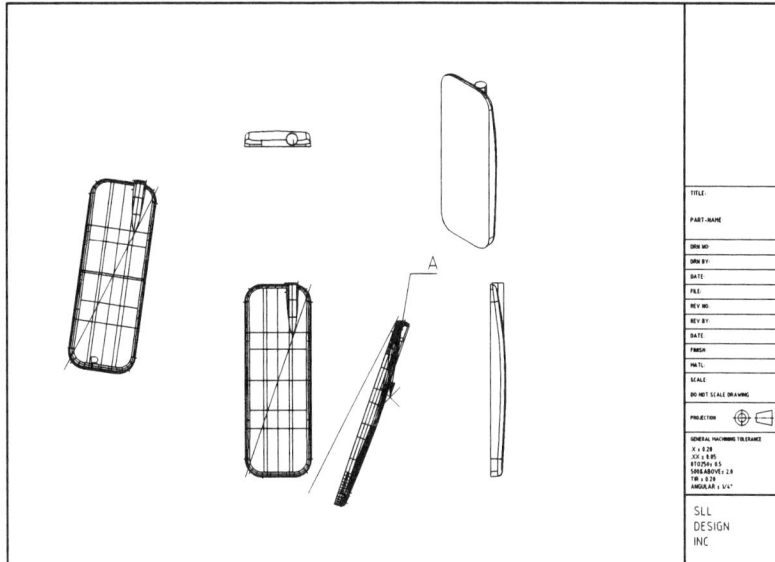

Figure 6.12 Second auxiliary view created

The opposite of creating a view is deleting a view. To delete a view, you use the AMDELVIEW command. Run this command to delete the first auxiliary view, which is the parent view of the second auxiliary view. If you delete a parent view, you can choose whether to delete its dependent view. See Figure 6.13.

<Drawings>   **<Edit View>**   **<Delete>**

Command: **AMDELVIEW**
Select view to delete: **[Select A (Figure 6.12).]**
View has 1 dependent view. Delete it also? Yes/No/<Cancel>: **YES**

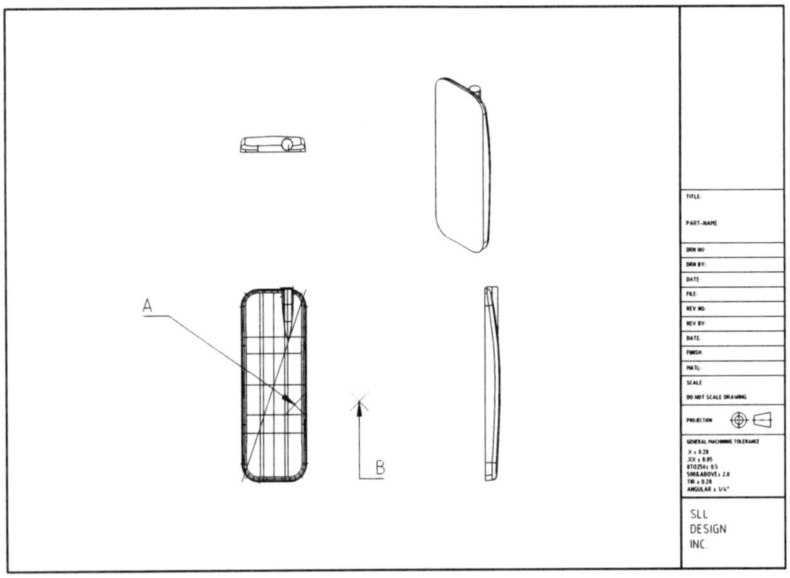

Figure 6.13 First auxiliary view and its dependent view deleted

As mentioned earlier, you can move a drawing view. Run the AMMOVEVIEW command to move the front view. When you move a parent view, all its dependent views move as well. See Figure 6.14.

<Drawings>   **<Edit View>**   **<Move>**

Command: **AMMOVEVIEW**
Select view to move: **[Select A (Figure 6.13).]**
3 descendant views will also be moved.
View location: **[Select B (Figure 6.13).]**
View location: **[Enter]**

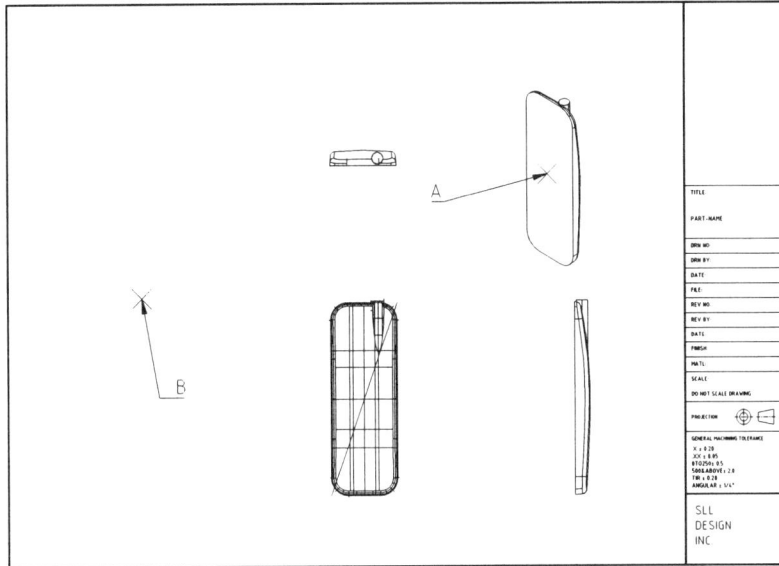

Figure 6.14 All views moved

Repeat the AMMOVEVIEW command to move the isometric view. This view does not have any dependent view. So, only this view moves. See Figure 6.15.

<Drawings>   **<Edit View>**   **<Move>**

Command: **AMMOVEVIEW**
Select view to move: **[Select A (Figure 6.14).]**
View location: **[Select B (Figure 6.14).]**
View location: **[Enter]**

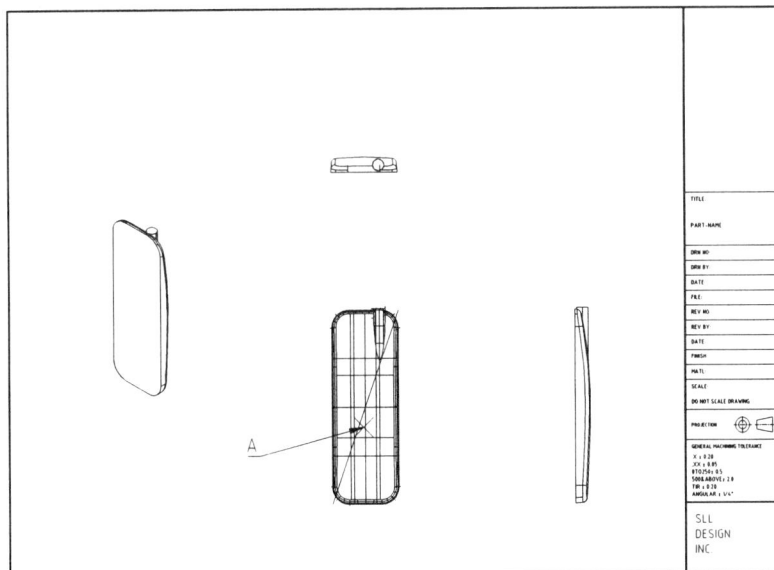

Figure 6.15 Isometric view moved

Earlier, for the purpose of displaying the reference line, you edited the front view to not calculate the hidden lines. To turn on hidden line calculation, you have to edit the view once more.

Run the AMEDITVIEW command again to edit the front view. See Figure 6.16.

<Drawings>   **<Edit View>**   **<Attributes>**

Command: **AMEDITVIEW**
Select view to edit: **[Select A (Figure 6.15).]**

> [Hidden Lines:
>   Calculate hidden lines   **Yes**
>   Hide hidden lines   **Yes**
>   **OK**   ]

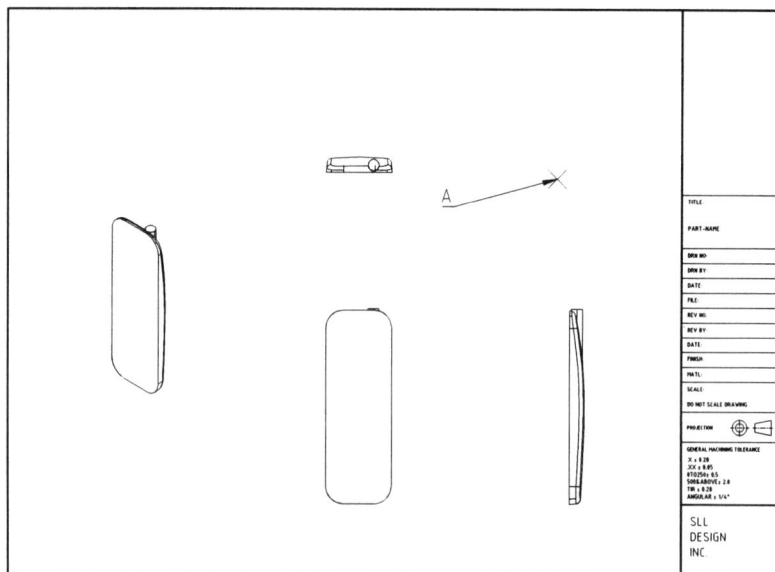

Figure 6.16 Front view edited

In the previous chapter, you have learned how to output flow lines, section lines, parting lines, and augmented lines from a surface model. If you want to show these lines on the drawing document, you must create a drawing view for this purpose and suppress hidden line calculation.

To illustrate how this can be done, switch back to model mode. Then, produce a section line.

Run the AMMODE command.

<Drawings>   **<Drawing Mode>**

Command: **AMMODE**
Drawing/<Model>: **[Enter]**

In model mode, rotate the UCS about the X axis for 90°. Then, set the current layer to UTY.

[UCS]   **[X Axis Rotate UCS]**

Command: **UCS**
Origin/ZAxis/3point/OBject/View/X/Y/Z/Prev/Restore/Save/Del/?/<World>: **X**
Rotation angle about X axis <0>: **90**

<Data>   <Layers...>

Command: **DDLMODES**

Current layer: **UTY**

After setting the UCS and the layer, create a set of section lines across the surface model. See Figure 6.17.

<Surfaces>   <Create Wireframe>   <Section Cuts...>

Command: **AMSECTION**
Select surfaces: **ALL**
Select surfaces: **[Enter]**

    [Section Type   **Single**
    Initial Plane   **UCS Plane**
    Save   **Yes**
    OK   ]

Use the GROUP command to put the section lines into an entity group called SECTION.

[Standard Toolbar] **[Object Group]**

Command: **GROUP**

    [Group name: **SECTION**
    Selectable: **Yes**
    New:]

Select objects for grouping:
Select objects: **[Select the newly created section lines.]**
Select objects: **[Enter]**

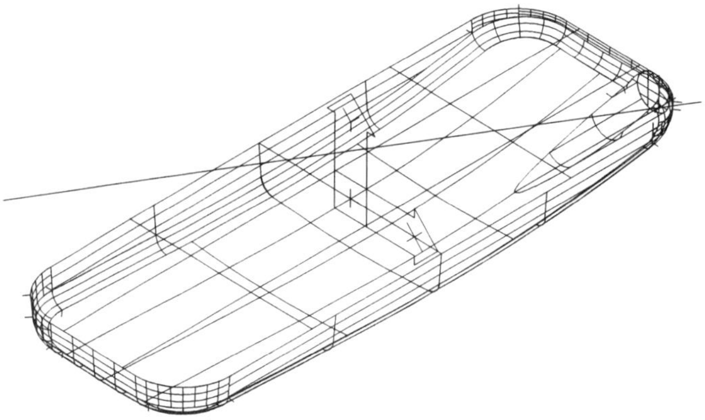

Figure 6.17 Cross section created

After producing a set of section lines across the surface model, return to drawing mode by running the AMMODE command.

<Drawings>   **<Drawing Mode>**

Command: **AMMODE**
Model/<Drawing>: **[Enter]**

To create a drawing view to display the section lines, use the AMDWGVIEW command. Because the section lines are not surfaces, you have to suppress hidden line calculation. See Figure 6.18.

<Drawings>   **<Create View>**

Command: **AMDWGVIEW**

    [Type: **Base**
     Hidden Lines:
            Calculate hidden lines   **No**
     Data Set: **Group**
     Scale: **0.5**
     OK   ]

Enter group name to be included in view: **SECTION**
worldXy/worldYz/worldZx/Ucs/<Select work plane or edge>: **UCS**
Rotate/Z-flip/<Accept>: **[Enter]**
Location for base view: **[Select A (Figure 6.16).]**
Location for base view: **[Enter]**

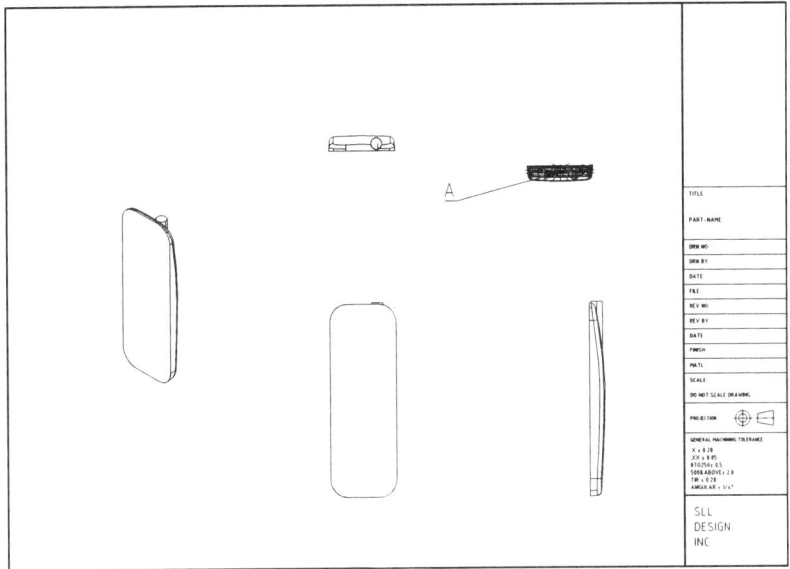

Figure 6.18 Base view added

As a result of suppressing hidden line calculation, you can see in Figure 6.18 that the section lines, together with the UV lines of the surfaces show up in the drawing. In order to hide the surface model in this drawing view, you can freeze the layer SURF in the current viewport.

To do so, run the MSPACE command to switch to floating model space.

<View>   <Floating Model Space>

Command: **MSPACE**

After switching to floating model space, select the new viewport to make it the active viewport. Next, run the DDLMODES command to freeze the layer SURF on the current viewport. See Figure 6.19.

<Data>   <Layers...>

Command: **DDLMODES**

Layer	Cur VP:
SURF	**Frz**

After freezing the layer SURF in the new viewport, return to paper space.

<View>   <Paper Space>

Command: **PSPACE**

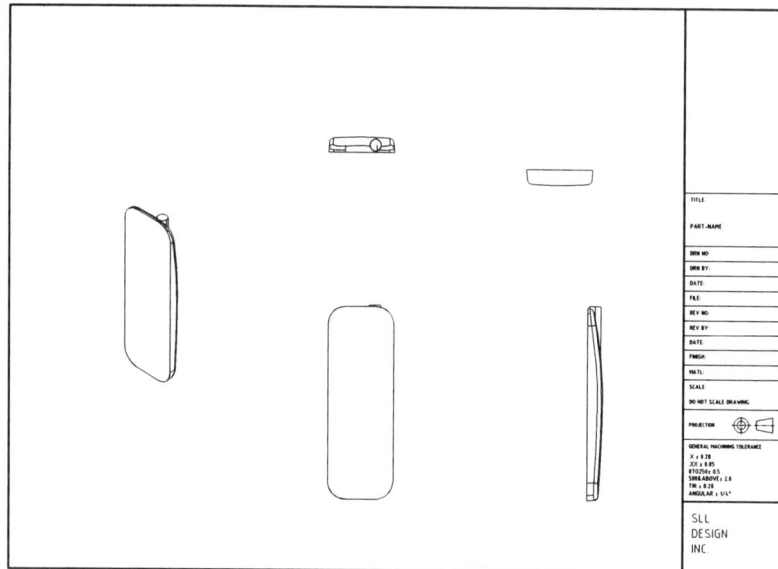

Figure 6.19 Layer SURF frozen in the current viewport of the new base view

Although it is common practice to add dimensions to an engineering document, you will not add dimensions to this drawing because the document shows a set of NURBS surfaces. As such, the displays on the drawing views are merely silhouettes of the surface model in a particular viewing direction. Dimensions set on these silhouettes do not seem to provide any real or valuable engineering meaning. To use a surface model to manufacture, it is strongly advised that you use electronic data transmission. To output a surface model, you can use the IGESOUT command or the AMMODOUT command.

You have completed the 2D engineering document for the surface model. This document contains a title block and several drawing views. The drawing views include a front view, a top view, a side view, and an isometric view.

You can retrieve information about a drawing view by using the AMLISTDWG command.

Command: **AMLISTDWG**
Select view: **[Select A (Figure 6.6).]**

Base Drawing View
id = 2   view is ACTIVE and up to date
view scale: 0.5000
view direction: 0.0000,0.0000,1.0000
center point: 111.6624,71.0882 target point : 0.0000,0.0000,10.0000
visible layer: AM_VIS
hidden layer: AM_HID   hidden layer linetype : HIDDEN
Hidden lines are blanked.
Tangent edges are not displayed.
View has 3 descendants, 0 dimensions, 0 notes.
Group containing 25 entities represented

With a properly installed plotting device, you can output a document of this model.

6.3 Summary

In this chapter, you used the following AutoSurf commands to create a 2D engineering drawing from a surface model.

AMDELVIEW	AMDWGVARS	AMDWGVIEW
AMEDITVIEW	AMLISTDWG	AMMODE
AMMOVEVIEW		

For a brief explanation of these commands, refer to the appendix of this book.

So far, you have learned how to create wireframes, how to put NURBS surfaces on the wireframes, how to use a NURBS surface model, and how to document a 3D model.

The appendix provides a brief summary of all the AutoSurf commands and variables.

6.4 Exercise

Follow the steps outlined in this chapter to open each of the drawing files you created in Chapters 3, 4, and 5. Then, switch to drawing mode and prepare an engineering document for each drawing.

Appendix
Quick Reference

In the previous chapters, you practiced creating NURBS wireframes and NURBS surface models, and learned what you can do with a surface model. By now, you should be able to use AutoSurf Release 3 as a tool to create 3D NURBS surface models that meet your design needs.

This appendix provides you with a summary of the AutoSurf Release 3 commands and variables as well as a table that lists the changes in command names and system variable names from AutoSurf Release 2.

A.1 Pull-Down Menus and Windows Toolbars

AutoSurf uses three additional pull-down menus and four additional toolbars.

The pull-down menus are
    Surfaces
    Drawings
    Mtools
The toolbars are
    Surface Create
    Surface Edit
    Drawing
    View

A.2 Surface Creation Commands

The following AutoSurf commands concern NURBS surfaces creation.

AM2SF	converts thick lines, arcs, circles and polylines, 3D polygon meshes, AutoCAD solids, and AutoCAD Designer solid to AutoSurf surfaces. You can convert selected faces of the object or the entire object into a set of surfaces.
AMBLEND	creates a surface by blending together two, three, or four wires and the edges or UV lines of two, three, or four surfaces.
AMCORNER	trims and rounds the corner of three intersecting fillet surfaces.
AMEXTRUDESF	extrudes lines, arcs, circles, ellipse, splines, or polylines to create an extruded surface.
AMFILLETSF	creates a constant or variable fillet surface from two intersecting surfaces.

AMLOFTU	interpolates a lofted U surface from a series of lines, arcs, polylines, splines, or augmented lines.
AMLOFTUV	interpolates a lofted UV surface from the contours of two sets of orthogonal lines, arcs, or splines.
AMOFFSETSF	derives an offset surface from a surface.
AMPLANE	creates a rectangular planar surface or a trimmed planar surface from a set of closed planar lines, arcs, splines, or polylines.
AMPRIMSF	creates primitive surfaces, such as cones, tori, spheres, and cylinders.
AMREVOLVESF	revolves lines, arcs, splines, or polylines about an axis to create a surface of revolution.
AMRULE	creates a straight element ruled surface between a pair of lines, arcs, splines, or polylines.
AMSWEEPSF	creates a swept surface by sweeping a set of lines, arcs, splines, or polylines along one or two rails; the first rail can be an augmented line.
AMTUBE	creates a tubular surface along an axis defined by a set of lines or polylines.

A.3   Surface Modification Commands

The following AutoSurf commands concern NURBS surfaces modification.

AMBREAK	breaks a single surface into a number of surfaces.
AMEDGE	copies and extracts the edges of a surface as a spline or polyline, untrims a trimmed surface, and temporary displays the nodes of the edges of a surface.
AMEDITSF	flips the normal direction of a surface, controls the span of influence of the grip points on the surface, truncates the base surface of a trimmed surface, and adjusts the number of grip points in U- and V directions.
AMINTERSF	trims one or both intersecting surfaces, and creates a polyline along the intersection of two surfaces.
AMJOINSF	joins surfaces at their untrimmed edges to form a larger surface.
AMLENGTHEN	extends or reduces the length of a surface at its untrimmed edges.
AMPROJECT	projects lines, circles, arcs, splines, or augmented lines onto a surface to create a wireframe or augmented line of the projected entities or to trim the surface based on the projected entities.
AMREFINESF	refines a surface by changing the number of U- and V-patches or the tolerance for fitting splines and surfaces.

A.4   3D Wireframe Creation and Editing Commands

The following AutoSurf commands concern wireframe creation and editing. You can use these commands in conjunction with other AutoCAD commands in wireframe creation.

AMAUGMENT	creates an augmented line from the edges, displayed UV lines, and trimmed edges of a surface.
AMDIRECTION	displays or reverses the direction of lines, arcs, splines, polylines, or augmented lines.
AMEDITAUG	edits the vectors of augmented line, and creates an augmented line from a line and polyline.
AMFILLET3D	creates a wireframe fillet between coplanar lines, arcs, splines, or polylines.
AMFITSPLINE	fits a smooth NURBS spline from selected polylines, lines, circles, ellipses, and arcs.
AMFLOW	creates 3D polylines or augmented lines along the U and V directions of a surface.
AMJOIN3D	joins lines, arcs, polylines, splines, and augmented lines to form a polyline, spline, or augmented line.
AMOFFSET3D	offsets a 3D polyline by a specified distance relative to the current view.
AMPARTLINE	creates a 3D parting line on a surface in the current view.
AMREFINE3D	refines a line or polyline by increasing or decreasing the number of control points.
AMSECTION	creates a section across a set of surfaces.
AMUNSPLINE	converts a spline to a polyline based on the control points.

A.5   Display Control Commands

The following AutoSurf commands concern display control.

AMDISPSF	controls the display mode of surfaces.
AMVIEW	controls the display of the current viewport.
AMVISIBLE	controls the visibility of individual entities.

A.6   Documentation Commands

The following AutoSurf commands concern documentation of NURBS surface models.

AMDELVIEW	deletes the selected view and its dependent views.
AMDWGVARS	sets the system variables for documentation.
AMDWGVIEW	creates a base, orthographic, auxiliary, isometric, or detail view in paper space. (Section view is available for AutoCAD Designer objects.)
AMEDITVIEW	edits an existing drawing view.
AMLISTDWG	lists information about a drawing view.
AMMODE	toggles between drawing mode and model mode.
AMMOVEVIEW	moves the position of drawing views.

A.7   Utility Commands

The followings are additional utility commands provided by AutoSurf.

AMCHECKFIT	checks the minimum 3D distance between lines, arcs, splines, polylines, or surface.
AMMODIN	imports files from AutoSurf/AutoMill Release 6.0 and Solution 3000 in .mod formats.
AMMODOUT	exports files to AutoSurf/AutoMill Release 6.0 and Solution 3000 in .mod formats.
AMSOLCUT	uses an AutoSurf surface to cut an AutoCAD solid or AutoCAD Designer solid.
AMSURFPROP	calculates the mass properties of AutoSurf surfaces.
AMSURFVARS	changes AutoSurf system variables.
AMVER	displays the version number of the AutoSurf software.

A.8   Related Shortcut Keys

The followings are useful shortcut keys for creating a surface model.

F	fits all the entities to the screen display.
S	starts the SPLINE command.
W	toggles between drawing mode and model mode.
QQ	edits the drawing view with the AMEDITVIEW command.
UU	sets the UCS to the current view.
VV	hides and unhides objects.
1	sets the display to a single viewport.
2	sets the display to two viewports.
3	sets the display to three viewports.
4	sets the display to four viewports.
5	sets the display to top view.
6	sets the display to front view.
7	sets the display to right view.
8	sets the display to isometric view.
9	sets the display to plan view of current UCS.
]	rotates the view to the right.
[	rotates the view to the left.
=	rotates the view upward.
-	rotates the view downward.

A.9   AutoSurf Variables

These are AutoSurf system variables.

AMBLENDTOL	controls the tolerance for blending surfaces and joining splines.
AMGRPREFIX	controls the default group prefix name.
AMJOINGAP	controls the default join gap distance.
AMPAGELEN	controls the text page length of the text window.

AMPFITANG	controls the default angle between the line segments of a polyline for fitting splines or surfaces.
AMPFITLEN	controls the default length of the line segments of a polyline for fitting splines or surfaces.
AMPFITTOL	controls the tolerance for fitting a spline to a polyline.
AMSFDISPMODE	controls the linetype for display surfaces.
AMSFTOL	controls the tolerance for surface accuracy.
AMULINES	controls the default number of display U flow lines on the surface.
AMVECAUG	controls the default length of the augmented vector.
AMVECSF	controls the default vector length of surfaces.
AMVERSION	stores the version number of AutoSurf.
AMVLINES	controls the default number of display V flow lines on the surface.

A.10 Related AutoCAD System Variables

The following AutoCAD system variables also concern AutoSurf.

CMDDIA	sets whether dialog box appears in a command.
DELOBJ	sets whether the original object is deleted after it has been used to create another object.

A.11 Command Name and System Variable Name Changes

If you have been using AutoSurf Release 2, you might notice that the AutoSurf command names in Release 3 have changed significantly and some commands are replaced by AutoCAD Release 13 commands. The following tables list the changes.

AutoSurf command name changes:

AutoSurf R2	AutoSurf R3	AutoCAD R13
ACAD2SF	AM2SF	———
ASURFVER	AMVER	———
BLENDSF	AMBLEND	———
BREAKSF	AMBREAK	———
BREAKVW	———	BREAK
CORNERSF	AMCORNER	———
CREATEAUG	AMAUGMENT	———
DIRECTION	AMDIRECTION	———
DISPLAY	AMVIEW	———
DISPSF	AMDISPSF	———
DIST3D	AMCHECKFIT	———
EDITAUG	AMEDITAUG	———
EDITBORDER	AMEDGE	———

AutoSurf R2	AutoSurf R3	AutoCAD R13
EDITSF	AMEDITSF	----
EDITSP	----	SPLINEDIT
ELLIPSE3D	----	ELLIPSE
ENTVIS	AMVISIBLE	----
EXPLODE3D	AMUNSPLINE	----
EXTENDSF	AMLENGTHEN	----
EXTENDVW	----	EXTEND
EXTRUDESF	AMEXTRUDESF	----
FILLET3D	AMFILLET3D	----
FILLETSF	AMFILLETSF	----
FLOWSF	AMFLOW	----
GROUPAS	----	GROUP
INTERSF	AMINTERSF	----
INTERVW	----	POINT
JOIN3D	AMJOIN3D	----
JOINSF	AMJOINSF	----
LINE2PT	----	----
LINEVW	----	LINE
LISTAS	----	LIST
LOFTSF	AMLOFTU	----
MESHSF	AMLOFTUV	----
MODIN	AMMODIN	----
MODOUT	AMMODOUT	----
NORMALSF	AMEDITSF	----
OFFSETSF	AMOFFSETSF	----
OFFSETVW	AMOFFSET3D	----
OUTPUTAUG	----	DXFOUT
PARTSF	AMPARTLINE	----
PLANESF	AMPLANE	----
PRIMSF	AMPRIMSF	----
PROFSF	----	----
PROJECT3D	----	----
PROJECTSF	AMPROJECT	----
PURGESF	----	----
REFINE3D	AMREFINE3D	----
REFINESF	AMREFINESF	----
REVOLVESF	AMREVOLVESF	----
RULESF	AMRULE	----
SECTSF	AMSECTION	----
SPLINE3D	AMFITSPLINE	SPLINE
SURFVAR	AMSURFVARS	----
SWEEPSF	AMSWEEPSF	----
TRIMVW	----	TRIM

AutoSurf R2	AutoSurf R3	AutoCAD R13
TRUNCSF	AMEDITSF	----
TUBESF	AMTUBE	----
UNSPLINE	AMUNSPLINE	----

AutoSurf system variable name changes:

AutoSurf R2	AutoSurf R3	AutoCAD R13
ASANGLE	AMPFITANG	----
ASAUGVECTOR	AMVECAUG	----
ASDEFORDER	----	----
ASDISPROT	----	----
ASFILLSFRAD	----	FILLETRAD
ASFLOWSU	----	----
ASFLOWSV	----	----
ASGRPPREFIX	AMGRPREFIX	----
ASJOINGAP	AMJOINGAP	----
ASLENGTH	AMPFITLEN	----
ASPAGELEN	AMPAGELEN	----
ASSAVEORIGINAL	----	DELOBJ
ASSPLINECONV	----	----
ASSURFDISP	----	----
ASSURFU	AMULINES	----
ASSURFV	AMVLINES	----
ASSURFVECTOR	AMVECSF	----
ASSYSTOL	AMSFTOL	----
ASSURFVER	AMSFVER	----
ASVARFILE	----	----

Index